SYNTHETIC TEXTILE

SYNTHETIC TEXTILE

By
Meenakshi Rastogi

SONALI

SONALI PUBLICATIONS
NEW DELHI-110002

First Published-2009
Reprinted, 2016
ISBN 978-81-8411-224-5

Published by
SONALI PUBLICATIONS
4228/1, Ansari Road,
New Delhi-110 002 (India)
Phone: 23266109, 23279245
Fax: 91-11-23253475
E-mail: sonalipub@yahoo.co.in

Printed at:
Sachin Printers, Delhi

Preface

Synthetic textile fibres have revolutionized the textile industry. Such artificial fibres are usually long-chain polymers, produced industrially by the condensation of many small units. Orlon is the trade name for a polyacrylonitrile fibre made from natural gas, oxygen, and nitrogen. It combines bulk with light weight and is resistant to acids and sun damage. It is used for sweaters and other clothing. Dacron is the trade name for a polyester fibre of great strength and wrinkle resistance. It is often blended with other fabrics. Vinyl fibres, such as Saran, are used for screening and heavy-duty upholstery.

The first artificial fibre, known as artificial silk from 1799 onwards, became known as Viscose around 1894, and finally Rayon in 1924. A similar product known as Cellulose Acetate was discovered in 1865. Rayon and Acetate are both artificial fibres, but not truly synthetic, being made from wood. Although these artificial fibres were discovered in the mid-nineteenth century, successful modern manufacture began much later (see the dates below).

Nylon, the first synthetic fibre, made its debut in the United States as a replacement for silk, just in time for World War II rationing. Its nobel use as a material for women's stockings overshadowed more practical uses, such as a replacement for the silk in parachutes and other military uses.

Synthetic (man-made) fabrics play a large role in our collection of technical apparel. Garments made with synthetic yarns are comfortable, adaptable, lightweight and dry quickly. Synthetic fleece keeps you warm and dry in the winter while synthetic base layer keeps you cool and dry in the summer. However, synthetic fabrics do require natural resources to produce. That's why whenever possible we source synthetic fabrics that contain recycled content.

Textile industry in India has started gaining prominence in the national agenda right from early years of the independence. India has vast picturesque locations, age-old cultures and traditions, architectural heritage as well as large and relatively inexpensive manpower resources (skilled and Unskilled) suitable for deployment in this industry; yet, these resources – natural and Manpower – have not been exploited optimally in India. There is enormous potential for growth in the industry leading to significant socio-economic benefits for the population.

"One hundred years of chemical fibres, and of these more than fifty years of synthetic fibres, do demand an inventory of the required technologies. While there is literature for the chemistry of the chemical and synthetic fibres, as well as for the processes and the textile testing technology, it is hardly possible to find comprehensive desciptions from installation, equipment and machinery manufacture through production of synthetic fibres and filaments and the manipulation of their properties..."

The objective of this book is to provide the whole knowledge about what kind of synthetic textiles are and in which form they exists, by which user can understand all the aspects of this book.

Thanks

Contents

1

Introduction

A material made mainly of natural or synthetic fibres. Modern textile products may be prepared from a number of combinations of fibres, yards, films, sheets, foams, furs, or leather. They are found in apparel, household and commercial furnishings, vehicles, and industrial products.

The term fabric may be defined as a thin, flexible material made of any combination of cloth, fibre, or polymer (film, sheet, or foams); cloth as a thin, flexible material made from yarns; yarn as a continuous strand of fibres; and fibre as a fine, rodlike object in which the length is greater than 100 times the diameter. The bulk of textile products are made from cloth.

The natural progression from raw material to finished product requires: the cultivation or manufacture of fibres; the twisting of fibres into yarns (spinning); the interlacing (weaving) or interloping (knitting) of yarns into cloth; and the finishing of cloth prior to sale.

The conversion of staple fibre into yarn (spinning) requires the following steps: picking (sorting, cleaning, and blending), carding and combing (separating and aligning), drawing (reblending), drafting (reblended fibres are drawn out into a long strand), and spinning (drafted fibres are further attenuated and twisted into yarn).

The process of weaving allows a set of yarns running in the machine direction (warp) to be interlaced with another set of yarns running across the machine (filling or weft). The weaving process involves four functions: shedding (raising the warp yarns by means of the appropriate harnesses); picking (inserting the weft yarn); battening (pushing the weft into the cloth with a reed); and taking up and letting off (winding the woven cloth onto the cloth beam and releasing more warp yarn from the warp beam;).

Newly constructed knit or woven fabric must pass through various finishing processes to make it suitable for its intended purpose. Finishing enhances the appearance of fabric and also adds to its serviceability. Finishes can be solely mechanical, solely chemical, or a combination of the two. Those finishes, such as scouring and bleaching, which simply prepare the fabric for further use are known as general finishes. Functional finishes, such as durable press treatments, impart special characteristics to the cloth.

Any filament, fibre, or yarn that can be made into fabric or cloth, and the resulting material itself. The word originally referred only to woven fabrics but now includes knitted, bonded, felted, and tufted fabrics as well. The basic raw materials used in textile production are fibres, either obtained from natural sources (e.g., wool) or produced from chemical substances (e.g., nylon and polyester). Textiles are used for wearing apparel, household linens and bedding, upholstery, draperies and curtains, wall coverings, rugs and carpets, and bookbindings, in addition to being used widely in industry.

A **textile** is a flexible material consisting of a network of natural or artificial fibres often referred to as thread or yarn. Yarn is produced by spinning raw wool fibres, linen, cotton, or other material on a spinning wheel to produce long strands known as yarn. Textiles are formed by weaving, knitting, crocheting, knotting, or pressing fibres together (felt).

The term 'Textile' is a Latin word originating from the word 'taxer' which means 'to weave'. Textile refers to a flexible material comprising of a network of natural or artificial fibres, known as yarn. Textiles are formed by weaving, knitting, crocheting, knotting and pressing fibres together. Textile Museum is that specialized category of museum which primarily preserves different types of textile and textile products.

The words *'fabric'* and *cloth* are used in textile assembly trades (such as tailoring and dressmaking) as synonyms for *textile*. However, there are subtle differences in these terms. *Textile* refers to any material made of interlacing fibres. *Fabric* refers to any material made through weaving, knitting, crocheting, or bonding. *Cloth* refers to a finished piece of fabric that can be used for a purpose such as covering a bed.

History of Textile

The History of textile is almost as old as that of human civilization and as time moves on, the history of textile has further enriched itself. In the 6th and 7th century BC, the oldest recorded indication of using fibre comes with the invention of flax and wool fabric at the excavation of Swiss lake inhabitants. In India the culture of silk was introduced in 400AD, while spinning of cotton traces back to 3000BC. In China, the discovery and consequent development of sericulture and spin silk methods got initiated at 2640 BC while in Egypt the art of spinning linen and weaving developed in 3400 BC. The discovery of machines and their widespread application in processing natural fibres was a direct outcome of the industrial revolution of the 18th and 19th centuries. The discoveries of various synthetic fibres like nylon created a wider market for textile products and gradually led to the invention of new and improved sources of natural fibre. The development of transportation and communication facilities facilitated the path of transaction of localized skills and textile art among various countries.

History of Various Countries' Textile

Indian Textile

Indian textile enjoys a rich heritage and the origin of textiles in India traces back to the Indus Valley Civilization where people used home spun cotton for weaving their clothes. Rigveda, the earliest of the Veda contains the literary information about textiles and it refers to weaving. Ramayana and Mahabharata, the eminent Indian epics depict the existence of wide variety of fabrics in ancient India. These epics refer both to rich and stylized garment worn by the aristocrats and ordinary simple clothes worn by the common people. The contemporary Indian textile not only reflects the splendid past but also cater to the requirements of the modern times.

Chinese Textile

Chinese textiles enjoy an excellent heritage in textile sector and occupy a prominent position in the global textile market. Chinese textiles are world famous and extraordinary for their fine quality and profound symbolic meanings. Textiles in China often form an integral aspect of its heritage and symbolically reflect its tradition and culture. In China, textile is often closely associated with prosperity and involved in the process of elaborate rituals. Parents' spontaneous love for their children is most visibly reflected through the excellent clothes they provide on festive occasions to their children. These clothes are made up of expensive materials and excellent craftsmanship.

African Textile

In the ancient times the most important aspect of textiles

or more precisely cloth in Africa was that cloth was used as form of money. The width of cloth strip was usually standardized in each region of Africa and therefore there used to be a regular number of such standard length cloth strips required to make a woman's wrapper cloth. This would then be used to serve as the unit of value. Cloth was a convenient form of money primarily because it was used by everybody, fairly durable and easily sub dividable.

The weavers, dyers and other textile artists of Africa together makes an active contribution in creating exquisite and amazing range of textiles. African usually embody a great variety of styles. Adinkara, kente and bogolan are some of the African textiles which are becoming increasingly popular while some others like Yoruba, Ase-oke and Adire are equally beautiful but less well known.

The textile industry, with its extremely long and rich history, has had a massive impact on the world economy and the very evolution of modern society. Weaving is believed to be one of the oldest surviving crafts in the world today, the actual origins of which are thought to date back to Neolithic times 12,000 years ago. Even before that time, the same principle was used to interlace branches and twigs to form protective fences, shelters and baskets. Once the practicality of interlacing these kinds of materials was understood, further experimentation with other natural materials probably produced the first basic fabrics and clothes.

Japanese Textile

In 1869, the capital of Japan was shifted from Kyoto to Tokyo and from this time onwards the Nishijin weaving tradition seemed threatened with extinction. The industry again started to grow along with Japan's new capitalist economy by 1890 when the Nishijin weavers embraced and applied modern technology to their own ancient and original textile art. The textile art of Japan particularly reached an epitome of excellence by exhibiting a cultural distinction and remarkable artistic skill in the Edo and succeeding Meiji periods (1868 - 1912).

Uses

Textiles have an assortment of uses, the most common of which are for clothing and containers such as bags and baskets. In the household, they are used in carpeting, upholstered furnishings, window shades, towels, covering for tables, beds, and other flat surfaces, and in art. In the workplace, they are used in industrial and scientific processes such as filtering. Miscellaneous uses include flags, backpacks, tents, nets, cleaning devices, such as handkerchiefs; transportation devices such as balloons, kites, sails, and parachutes; and strengthening in composite materials such as fibre glass and industrial geotextiles. Textiles can be used for educational purposes. Textiles can be used as a material for children to use and explore in their classrooms as another element of learning. Children can manipulate and come up with creative uses for textiles such as collage materials, art materials and so on.

Textiles used for industrial purposes, and chosen for characteristics other than their appearance, are commonly referred to as *technical textiles.* Technical textiles include textile structures for automotive applications, medical textiles (e.g. implants), geotextiles (reinforcement of embankments), agro textiles (textiles for crop protection), protective clothing (e.g. against heat and radiation for fire fighter clothing, against molten metals for welders, stab protection, and bullet proof vests. In all these applications stringent performance requirements must be met. Woven of threads coated with zinc oxide nanowires, laboratory fabric has been shown capable of "self-powering nanosystems" using vibrations created by everyday actions like wind or body movements.

Sources and Types

Textiles can be made from many materials. These materials come from four main sources: animal, plant, mineral, and synthetic. In the past, all textiles were made from natural fibres, including plant, animal, and mineral sources. In the 20th century, these were supplemented by artificial fibres made from petroleum.

Textiles are made in various strengths and degrees of durability, from the finest gossamer to the sturdiest canvas. The relative thickness of fibres in cloth is measured in deniers. Micro fibre refers to fibres made of strands thinner than one denier.

Animal Textiles

Animal textiles are commonly made from hair or fur. Wool refers to the hair of the domestic goat or sheep, which is distinguished from other types of animal hair. That the individual strands are coated with scales and tightly crimped, and the wool as a whole is coated with an oil known as lanolin, which is waterproof and dirtproof. Woollen refers to a bulkier yarn produced from carded, non-parallel fibre, while worsted refers to a finer yarn which is spun from longer fibres which have been combed to be parallel. Wool is commonly used for warm clothing. Cashmere, the hair of the Indian Cashmere goat, and Mohair, the hair of the North African angora goat, are types of wool known for their softness.

Other animal textiles which are made from hair or fur are alpaca wool, vicuna wool, llama wool, and camel hair, generally used in the production of coats, jackets, ponchos, blankets, and other warm coverings. Angora refers to the long, thick, soft hair of the angora rabbit.

Wadmal is a coarse cloth made of wool, produced in Scandinavia, mostly 1000-1500.

Silk is an animal textile made from the fibres of the cocoon of the Chinese silkworm. This is spun into a smooth, shiny fabric prized for its sleek texture.

Plant Textiles

Grass, rush, hemp, and sisal are all used in making rope. In the first two, the entire plant is used for this purpose, while in the last two, only fibres from the plant are utilized. Coir (coconut fibre) is used in making twine, and also in floormats, doormats, brushes, mattresses, floor tiles, and sacking.

Straw and bamboo are both used to make hats. Straw, a dried form of grass, is also used for stuffing, as is kapok.

Fibres from pulpwood trees, cotton, rice, hemp, and nettle are used in making paper.

Cotton, flax, jute, hemp and modal are all used in clothing. Pina (pineapple fibre) and ramie are also fibres used in clothing, generally with a blend of other fabrics such as cotton.

Acetate is used to increase the shininess of certain fabrics such as silks, velvets, and taffetas.

Seaweed is used in the production of textiles. A water-soluble fibre known as Alginate is produced and is used as a holding fibre; when the cloth is finished, the alginate is dissolved, leaving an open area

Tencel is a man-made fabric derived from wood pulp. It is often described as a man-made silk equivalent and is a tough fabric which is often blended with other fabrics - cotton for example.

Mineral Textiles

Asbestos and basalt fibre are used for vinyl tiles, sheeting, and adhesives, “transit” panels and siding, acoustical ceilings, stage curtains, and fire blankets.

Glass Fibre is used in the production of spacesuits, ironing board and mattress covers, ropes and cables, reinforcement fibre for composite materials, insect netting, flame-retardant and protective fabric, soundproof, fireproof, and insulating fibres.

Metal fibre, metal foil, and metal wire have a variety of uses, including the production of cloth-of-gold and Jewellery.

Hardware cloth is a coarse weave of steel wire, used in construction.

Synthetic Textiles

A variety of contemporary fabrics. From the left: even weave cotton, velvet, printed cotton, calico, felt, satin, silk, Hessian, polycotton. All synthetic textiles are used primarily in the production of clothing.

Polyester fibre is used in all types of clothing, either alone or blended with fibres such as cotton.

Aramid fibre (e.g., Twaron) is used for flame-retardant clothing, cut-protection, and armor.

Acrylic is a fibre used to imitate wools, including cashmere, and is often used in replacement of them.

Nylon is a fibre used to imitate silk; it is used in the production of pantyhose. Thicker nylon fibres are used in rope and outdoor clothing.

Spandex (trade name *Lycra*) is a polyurethane fibre that stretches easily and can be made tight-fitting without impeding movement. It is used to make active wear and swimsuits.

Olefin fibre is a fibre used in active wear, linings, and warm clothing. Olefins are hydrophobic, allowing them to dry quickly. A sintered felt of olefin fibres is sold under the trade name Tyvek.

Ingeo is a polylactide fibre blended with other fibres such as cotton and used in clothing. It is more hydrophilic than most other synthetics, allowing it to wick away perspiration.

Lurex is a metallic fibre used in clothing embellishment.

Types of Fibre

Fibres is a class of hair-like materials that are continuous filaments or are in discrete elongated pieces, similar to pieces of thread. They can be spun into filaments, thread, or rope. They can be used as a component of composite materials.

They can also be matted into sheets to make products such as paper or felt. Fibres are of two types: **natural fibre** and **man made** or **synthetic fibre.**

Natural Fibres

Natural fibres include those made from plant, animal and mineral sources. Natural fibres can be classified according to their origin.

Vegetable Fibres

Vegetable fibres are generally comprised mainly of cellulose: examples include cotton, jute, flax, ramie, sisal, and hemp. Cellulose fibres serve in the manufacture of paper and cloth.

COTTON- Cotton is a soft, staple fibre that grows in a form known as a boll around the seeds of the cotton plant (*Gossypium* sp.), a shrub native to tropical and subtropical regions around the world, including the America, India and Africa. The fibre most often is spun into yarn or thread and used to make a soft, breathable textile, which is the most widely used natural-fibre cloth in clothing today.

JUTE-Jute is a long, soft, shiny vegetable fibre that can be spun into coarse, strong threads. It is produced from plants in the genus *Corchorus*, family Tiliaceae.

Jute is one of the cheapest natural fibres and is second only to cotton in amount produced and variety of uses. Jute fibres are composed primarily of the plant materials cellulose (major component of plant fibre) and lignin (major components wood fibre). It is thus a ligno-cellulosic fibre that is partially a textile fibre and partially wood. It falls into the bast fibre category (fibre collected from bast or skin of the plant) along with kenaf, industrial hemp, flax (linen), ramie, etc. The industrial term for jute fibre is *raw jute*. The fibres are off-white to brown, and 1–4 meters (3–12 feet) long.

FLAX- Flax (also known as **common flax** or **linseed**) (binomial name: ***Linum usitatissimum***) is a member of the genus *Linum* in the family Linaceae. It is native to the region extending from the eastern Mediterranean to India and was probably first domesticated in the Fertile Crescent. This is called as Jawas/Javas or Alashi in Marathi. Flax was extensively cultivated in ancient Egypt. (NewZealand flax is not related to flax, but was named after it as both plants are used to produce fibres.)

Flax is an erect annual plant growing to 1.2 m tall, with slender stems. The leaves are glaucous green, slender lanceolate, 20–40 mm long and 3 mm broad. The flowers are pure pale blue, 15–25 mm diameter, with five petals; they can also be bright red. The fruit is a round, dry capsule 5–9 mm diameter, containing several glossy brown seeds shaped like an apple pip, 4–7 mm long.

Uses- Flax is grown both for its seeds and for its fibres. Various parts of the plant have been used to make fabric, dye, paper, medicines, fishing nets and soap. It is also grown as an ornamental plant in gardens.

RAMIE- Ramie (*Boehmeria nivea*) is a flowering plant in the nettle family Urticaceae, native to eastern Asia. It is a herbaceous perennial growing to 1 - 2.5 m tall; the leaves are heart-shaped, 7-15 cm long and 6-12 cm broad, and white on the underside with dense small hairs - this gives it a silvery appearance; unlike nettles, the hairs do not sting. The true Ramie or China Grass also called Chinese plant or white Ramie is the Chinese cultivated plant. A second type, is known as green Ramie or Rhea and is believed to have originated in the Malaya Peninsula. This type has smaller leaves which are green on the underside, and it appears to be better suited to tropical conditions.

SISAL- Sisal is an agave. *Agave sisalana* that yields a stiff fibre traditionally used in making twine and rope. (The term may refer either to the plant or the fibre, depending on context.) It is sometimes been incorrectly referred to as *sisal hemp* because hemp was for centuries a major source of fibre, so other fibres were sometimes named after it.

The plant originated in the Yucatan Peninsula of Mexico. In the 19th century, Sisal cultivation spread to Florida, the Caribbean islands and Brazil, as well as to countries in Africa, notably Tanzania and Kenya, and Asia. The first commercial plantings in Brazil were made in the late 1930's and the first Sisal fibre exports from there were made in 1948. It was not until the 1960,s that Brazilian production really accelerated and the first of many spinning mills were established. Today, Brazil is the major world producer of Sisal. There are both positive and negative environmental impacts from Sisal growing. Traditionally used for rope and twine, sisal has many uses, including paper, cloth, wall coverings and carpets.

HEMP- Hemp (from Old English *hænep*, see cannabis (etymology)) is the common name for plants of the entire genus *Cannabis*, although the term is often used to refer only to *Cannabis* strains cultivated for industrial (non-drug) use.

Industrial Hemp has many uses, including paper, textiles, biodegradable plastics, health food, and fuel. It is one of the fastest growing biomasses known, and one of the earliest domesticated plants known. It also runs parallel with the "Green Future" objectives that are becoming increasingly popular. Hemp requires little to no pesticides, no herbicides, controls erosion of the topsoil, and produces oxygen. Furthermore, Hemp can be used to replace many potentially harmful products, such as tree paper (the processing of which uses bleaches and other toxic chemicals, and contributes to deforestation), cosmetics, and plastics, most of which are petroleum-based and do not decompose easily

This fibre can be further categorized into the following:

- Seed fibre: Fibres collected from seeds or seed cases. e.g., Cotton and Kapok.
- Leaf fibre: Fibres collected from leaves. e.g. sisal and agave.
- Bast fibre or skin fibre: Fibres are collected from the skin or bast surrounding the stem of their respective plant.

These fibres have higher tensile strength than other fibres. Therefore, these fibres are used for durable yarn, fabric, packaging, and paper. Some examples are flax, jute, kenaf, industrial hemp, ramie, rattan, soybean fibre, and even vine fibres and banana fibres.

- Fruit fibre: Fibres are collected from the fruit of the plant, e.g., coconut (coir) fibre.
- Stalk fibre: Fibres are actually the stalks of the plant. E.g., straws of wheat, rice, barley, and other crops including bamboo and grass. Tree wood is also such a fibre.

The most used natural fibres are cotton, flax and hemp, although sisal, jute, kenaf, and coconut are also widely used.

Hemp fibres are mainly used for ropes and aerofoils because of their high suppleness and resistance within an aggressive environment. Hemp fibres are, for example, currently used as a seal within the heating and sanitary industries.

Animal fibres : Animal fibres generally comprise proteins; examples include silk, wool, angora, mohair and alpaca.

- Animal hair (wool or hairs): Fibre or wool taken from animals or hairy mammals. e.g. sheep's wool, goat hair (cashmere, mohair), alpaca hair, horse hair, etc.
- Silk fibre: Fibre collected from dried saliva of bugs or insects during the preparation of cocoons. Examples include silk.
- Avian fibre: Fibres from birds, e.g. feathers and feather fibre.

Synthetic Fibre

Synthetic fibres are the result of extensive research by scientists to improve upon naturally occurring animal and plant. In general, synthetic fibres are created by forcing, usually through extrusion, fibre forming materials through holes (called spinnerets) into the air, forming a thread. Before Synthetic fibres were developed, artificially manufactured fibres were made from cellulose, which comes from plants.

Synthetic Textile Fibres : have revolutionized the textile industry. Such artificial fibres are usually long-chain polymers, produced industrially by the condensation of many small units. Orlon is the trade name for a polyacrylonitrile fibre made from natural gas, oxygen, and nitrogen. It combines bulk with light weight and is resistant to acids and sun damage. It is used for sweaters and other clothing. Dacron is the trade name for a polyester fibre of great strength and wrinkle resistance. It is often blended with other fabrics. Vinyl fibres, such as Saran, are used for screening and heavy-duty upholstery.

The first artificial fibre, known as artificial silk from 1799 onwards, became known as Viscose around 1894, and finally Rayon in 1924. A similar product known as Cellulose Acetate was discovered in 1865. Rayon and Acetate are both artificial fibres, but not truly synthetic, being made from wood. Although these artificial fibres were discovered in the mid-nineteenth century, successful modern manufacture began much later (see the dates below).

Nylon, the first synthetic fibre, made its debut in the United States as a replacement for silk, just in time for World War II rationing. Its nobel use as a material for women's stockings overshadowed more practical uses, such as a replacement for the silk in parachutes and other military uses.

Common Synthetic fibres include:

- Rayon (1910) (artificial, not synthetic)
- Acetate (1924) (artificial, not synthetic)
- Nylon (1939)
- Modacrylic (1949)
- Olefin (1949)
- Acrylic (ORLON)(1950)
- Polyester (1953)
- PLA (2002)

Speciality Synthetic fibres include:

- Vinyon (1939)

- Saran (1941)
- Spandex (1959)
- Vinalon (1939)
- Aramids (1961)-known as Nomex, Kevlar and Twaron
- PBI (Polybenzimidazole fibre) (1983)
- Sulfar (1983)
- Lyocell (1992)
- Zylon (PBO fibre)

Modern fibres that are made from older artificial materials include:

- Glass Fibre is used for:
  - industrial, automotive, and home insulation (Fibre glass)
  - reinforcement of composite and plastics
  - specialty papers in battery separators and filtration
- Metallic fibre (1946) is used for:
  - adding metallic properties to clothing for the purpose of fashion (usually made with composite plastic and metal foils)
  - elimination and prevention of static charge build-up
  - conducting electricity to transmit information
  - conduction of heat

➛ Rayon is a manufactured regenerated cellulosic fibre. Because it is produced from naturally occurring polymers, it is neither a truly synthetic fibre nor a natural fibre; it is a semi-synthetic fibre. Rayon is known by the names *viscose rayon* and *art silk* in the textile industry. It usually has a high lustre quality giving it a bright shine. Rayon contains the chemical elements carbon, hydrogen, and oxygen.

➛ **Cellulose acetate,** first prepared in 1865, is the acetate ester of cellulose. Cellulose acetate is used as a film base in photography, and as a component in some adhesives; it is also used as a Synthetic fibre

Acetate Fibre and Triacetate Fibre

Acetate and Triacetate are mistakenly referred to as the same fibre; although they are similar, their chemical compounds differ. Triacetate is known as a generic description or primary acetate containing no hydroxyl group. Acetate fibre is known as modified or secondary acetate having a few hydroxyl groups. Triacetate fibres, although no longer produced in the United States, contain a higher ratio of acetate-to-cellulose than to acetate fibres.

➛ **Nylon** is a generic designation for a family of synthetic polymers known generically as polyamides and first produced on February 28, 1935 by Wallace Carothers at Dopant. Nylon is one of the most commonly used polymers.

Nylon is a thermoplastic silky material, first used commercially in a nylon-bristled toothbrush (1938), followed more famously by women's stockings ("nylons"; 1940). It is made of repeating units linked by peptide bonds (another name for amide bonds) and is frequently referred to as *polyamide* (PA). Nylon was the first commercially successful polymer and the first synthetic fibre to be made entirely from coal, water, and air. These are formed into monomers of intermediate molecular weight, which are then reacted to form long polymer chains.

Nylon was intended to be a synthetic replacement for silk and substituted for it in many different products after silk became scarce during World War II. It replaced silk in military applications such as parachutes and flak vests, and was used in many types of vehicle tires.

Nylon fibres are used in many applications, including fabrics, bridal veils, carpets, musical strings, and rope.

Solid nylon is used for mechanical parts such as machine screws, gears and other low-to medium-stress components previously cast in metal. Engineering-grade nylon is processed by extrusion, casting, and injection molding. Solid nylon is used in hair combs. Type 6/6 Nylon 101 is the most common commercial grade of nylon, and Nylon 6 is the most common commercial grade of molded nylon. Nylon is available in glass-filled variants which increase structural and impact strength and rigidity, and molybdenum sulfide-filled variants which increase lubricity.

Aramids are another type of polyamide with quite different chain structures which include aromatic groups in the main chain. Such polymers make excellent ballistic fibres.

Nylon Fibre

The Federal Trade Commission's definition for Nylon Fibre: "A manufactured fibre in which the fibre forming substance is a long-chain synthetic polyamide in which less than 85% of the amide-linkages are attached directly (-CO-NH-) to two aliphatic groups."

- A synthetic thermoplastic fibre (Nylon melts/glazes easily at relatively low temperatures)
- Round, smooth, and shiny filament fibres
- Cross sections can be either
  - o trilobal to imitate silk
  - o multipole to increase staple like appearance and hand
- Its most widely used structures are multifilament, monofilament, staple or tow and is available as partially drawn or as finished filaments.
- Regular nylon has a round cross section and is perfectly uniform. The filaments are generally completely transparent unless they have been delustered or solution dyed. Thus, they are microscopically recognized as glass rods.
- Molecular chains of nylon are long and straight variations but have no side chains or linkages.
  - o Cold drawing can align the chains so they are oriented with the lengthwise direction and are highly crystalline.

- Nylon is related chemically to the protein fibres, silk and wool.
    - They both have similar dye sites but nylon has many fewer dye sites than wool.

➛ **A Modacrylic** is a synthetic co-polymer. Modacrylics are soft, strong, resilient, and dimensionally stable. They can be easily dyed, show good press and shape retention, and are quick to dry. They have outstanding resistance to chemicals and solvents, are not attacked by moths or mildew, and are nonallergenic. Among their uses are in apparel linings, furlike outerwear, paint-roller covers, scatter rugs, carpets, and work clothing and as hair in wigs.

Commercial production of Modacrylic fibre began in 1949 by Union Carbide Corporation in the United States . Modacrylic and acrylic fibres are similar in composition and at one time were in the same category. In 1960 the Federal Trade Commission decided to separate the two fibres and establish a category for each.

The Federal Trade Commission defines Modacrylic fibres as manufactured fibres in which the fibre-forming substance is any long-chain synthetic polymer composed of less than 85%, but at least 35% weight of acrylonitrile units except when the polymer qualifies as rubber.

Modacrylic fibres are modified acrylic fibres made from acrylonitriles, but larger amounts of other polymers are added to make the copolymers.

The Modacrylic fibres are produced by polymerizing the components, dissolving the copolymer in acetone, pumping the solution into the column of warm air (dry-spun), and stretching while it is hot.

Modacrylics are creamy or white and are produced in tow and staple form. If looked at in cross section views, they have an irregular shape. Modacrylic fibres are also produced in many different lengths, crimp levels, deniers and they can have various shrinkage potentials.

Current modacrylic fibre producers include Solutia Inc. in the U.S. and Kaneka Corporation in Japan.

***Properties* :** A Modacrylic has properties that are similar to an Acrylic. However, modacrylics are flame retardant and do not combust. The fibres are difficult to ignite and are self-extinguish. In addition to a modacrylic's flame retardant properties it has a relatively high durability that is comparable to wool. Modacrylic fibres have a moderate resistance to abrasion and a very low tenacity.

Modacrylics are poor conductors of heat. The fabrics are soft, warm and resilient but are prone to pilling and matting. Modacrylics display high performance when it comes to appearance retention. The fibres are quite resilient and will not wrinkle. They also have great dimensional stability and high elastic recovery, which gives them the ability to hold their shape.

***Care* :** Modacrylics are sensitive to loss of appearance due to improper care; therefore, it is important to know how to care for modacrylics. Modacrylics are resistant to acids, weak alkalis, and organic solvents. These fibres are also resistant to moths, mildew and sunlight. Modacrylic fabrics can be machine washed using warm water and tumble dried on a low setting. Modacrylic pieces can also be dry-cleaned, however, they should not be steamed and should only be tumbled on cold. Some fabrics may also be cleaned using the furrier method (a special non-immersion cleaning process). The fibres are heat sensitive and will shrink at 250 °F and will stiffen at temperatures over 300 °F.

***Uses* :** Modacrylics are used primarily in applications where environmental resistance or flame retardancy is necessary or required. Modacrylics have the ability to combine flame retardancy with a relatively low density, meaning protective gear is not uncomfortably heavy (i.e. shirts and trousers worn by electrical linemen). The combination of flame retardancy and low density is also useful in furnishings, draperies, and outdoor fabrics.

Modacrylics are also commonly used in furlike fabrics, hairpieces, wigs and fleece type fabric. By mixing the various forms of fibres it is easy to create a realistic fur look. The fabrics can then be sheared or embossed to create a closer resemblance to fur. The heat-sensitivity of modacrylic also allows wigs and hairpiece to be curled and heat styled without damage.

Modacrylics are also used in fleece, knit-pile fabric backings, and non-woven fabrics. Other uses of modacrylics include paint rollers, industrial fabrics, stuffed toys and filters.

➛ **Olefin fibre** is a synthetic fibre made from alkenes. It is used in the manufacture of various textiles as well as clothing, upholstery, wallpaper, ropes, and vehicle interiors. Olefin is also referred to as polypropylene, polyethylene, or polyolefin. Olefin's advantages are its strength, colorfastness and comfort, stain, mildew, abrasion and sunlight resistance, and good bulk and cover.

Olefin fibres have great bulk and cover while having low specific gravity. This means "warmth without the weight." The fibres have low moisture absorption, but they can wick moisture and dry quickly. Olefin is abrasion, stain, sunlight, and chemical resistant. It does not dye well, but has the advantage of being colour fast. Since Olefin has a low melting point, textiles can be thermally bonded. The fibres have the lowest static of all manufactured fibres and a medium lustre. One of the most important properties of Olefin is its strength. It keeps its strength in wet or dry conditions and is very resilient. The fibre can be produced for strength of different properties.

Physical and Chemical Structure

Physical : Olefin fibres can be multi or monofilament and staple, tow or film yarns. The fibres are colourless and round in cross section. This cross section can be modified for different end uses. The physical characteristics are a waxy feel and colourless.

***Chemical* :** There are two types of polymers that can be used in Olefin fibres. The first, Polyethylene, is a simple linear structure with repeating units. These fibres are used mainly for ropes, twines and utility fabrics.

The second type, Polypropylene, is a three dimensional structure with a backbone of carbon atoms. Methyl groups protrude from this backbone. Stereos elective polymerization orders these methyl groups to the same spatial placement. This creates a crystalline polypropylene polymer. The fibres made with these polymers can be used in apparel, furnishing and industrial products.

➤ **Acrylic fibres** are synthetic fibres made from a polymer (Polyacrylonitrile) with an average molecular weight of 100,000, about 1900 monomer units. To be called Acrylic in the U.S, the polymer must contain at least 85% acrylonitrile monomer. Typical comonomers are vinyl acetate or methyl acrylate. The Dupont Corporation created the first acrylic fibres in 1941 and trademarked them under the name "**Orlon**".

Acrylic is lightweight, soft, and warm, with a wool-like feel. It dyes very well and has excellent colorfastness. It is resilient, retains its shape, and resists shrinkage and wrinkles. Acrylic has recently been used in clothing as a less expensive alternative to cashmere, due to the similar feeling of the materials. The disadvantages of Acrylic are that it tends to fuzz (or pill) easily and that it does not insulate the wearer as well as cashmere. Many products like fake pashmina or cashmina use this fibre to create the illusion of cashmere to the consumer. Acrylic is resistant to moths, oils, and chemicals, and is very resistant to deterioration from sunlight exposure.

➤ **Polyester** is a category of polymers which contain the ester functional group in their main chain. Although there are many polyesters, the term "polyester" as a specific material most commonly refers to Polyethylene Terephthalate (PET). Polyesters include naturally-occurring chemicals, such as in the cutin of plant cuticles, as well as synthetics such as polycarbonate and poly butyrate. Polyesters aren't biodegradable.

Polyesters may be produced in numerous forms such as sheets and three-dimensional shapes. Polyesters as thermoplastics may change shape after the application of heat. While combustible at high temperatures, polyesters tend to shrink away from flames and self-extinguish upon ignition. Polyester fibres have high tenacity and E-modulus as well as low water absorption and minimal shrinkage in comparison with other industrial fibres.

Woven polyester fabrics are used in consumer apparel and home furnishings such as bed sheets, bedspreads, curtains and draperies. Similarly, industrial polyesters are used in tyre reinforcements, ropes, fabrics for conveyor belts, safety belts, coated fabrics and plastic reinforcements with high energy absorption. Polyester fibre fills are also used to stuff pillows, comforters and cushion padding.

Polyester fabrics are claimed to have a "less natural" feel when compared to similarly-woven fabrics made from natural fibres (i.e. cotton in textile uses). However, polyester fabrics may exhibit other advantages over natural fabrics, such as improved wrinkle resistance. As a result, polyester fibres are sometimes spun together with natural fibres to produce a cloth with blended properties.

Polyesters are also used to make bottles, films, tarpaulin, canoes, liquid crystal displays, holograms, filters, dielectric film for capacitors, film insulation for wire and insulating tapes.

Liquid crystalline polyesters are among the first industrially-used liquid crystalline polymers. They are used for their mechanical properties and heat-resistance. These traits are also important in their application as an abradable seal in jet engines.

Thermosetting polyesters are used as casting materials, and Chemosetting polyester resins are used as fibre glass laminating resins and non-metallic auto-body fillers. Fibre glass-reinforced unsaturated polyesters find wide application in bodies of yachts and as body parts of cars.

Polyesters are also widely used as a finish on high-quality wood products such as guitars, pianos and vehicle / yacht interiors. Cured polyesters can be sanded and polished to a high-gloss, durable finish.

Mechanical Properties

Energy absorption of chemical fibre reinforced plastics (impact, bending and tensile tests) Investigation of the practical requirements for measuring the energy absorption of composite materials, and development of a suitable method for carrying out such measurements. A number of dynamic testing methods for measuring the energy absorption of laminates are reviewed, including an impact bending test, repeated-impact tests, an impact tensile test, and a ram bending test. Also discussed are impact tests on plate laminates. Particular emphasis is placed in these studies on composites with a chemical fibre reinforcement. It is established that a relation exists between the quasi-static energy absorption of the fibres and the dynamic energy absorption of the composite. Composites with commercial polyester and polyamide fibres lead to the highest energy absorptions, in which the testing apparatus has a significant effect.

➛ **Polylactic acid** or **Polylactide (PLA)** is a biodegradable, thermoplastic, aliphatic polyester derived from renewable resources, such as corn starch (in the U.S.) or sugarcanes (in rest of the world). Although PLA has been known for more than a century, it has only been of commercial interest in recent years, in light of its biodegradability.

Chemical and Physical Properties

Due to the chiral nature of lactic acid, several distinct forms of polylactide exist: poly-L-lactide (**PLLA**) is the product resulting from polymerization of L,L-lactide (also known as L-lactide). PLLA has a crystallinity of around 37%, a glass transition temperature between 50-80 °C and a melting temperature between 173-178 °C.

Polylactic acid can be processed like most thermoplastics into fibre (for example using conventional melt spinning processes) and film. The melting temperature of PLLA can be increased 40-50 °C and its heat deflection temperature can be increased from approximately 60°C to up to 190 °C by physically blending the polymer with PDLA (poly-D-lactide). PDLA and PLLA form a highly regular stereo complex with increased crystallinity. The temperature stability is maximised when a 50:50 blend is used, but even at lower concentrations of 3-10% of PDLA, there is still a substantial improvement. In the latter case, PDLA acts as a nucleating agent, thereby increasing the crystallization rate. Biodegradation of PDLA is slower than for PLA due to the higher crystallinity of PDLA. PDLA has the useful property of being optically transparent.

➛ **Vinyon** is a synthetic fibre made from Polyvinyl Chloride. In some countries other than the United States, Vinyon fibres are referred to as **polyvinyl chloride fibres**. It can bind non-woven fibres and fabrics. It was invented in 1939.

It has the same health problems associated with chlorinated polymers. In the past, Vinyon was used as a substitute for plant-based filters in tea bags.

Vinyon Fibre Characteristics

- Doesn't flame, but softens at low temperatures (55C)
- High resistance to chemicals
- Moisture absorption is less than 0.5% and moisture regained is less than 0.1%

Major vinyon fibre uses :

- Industrial applications as a bonding agent for non-woven fabrics and products.

➛ **Saran** is the trade name for a number of polymers made from Vinylidene Chloride (especially polyvinylidene chloride or **PVDC**), along with other monomers. The main advantage of Saran film is a very low permeability to water vapour, flavour and aroma molecules, and oxygen compared to other plastics. The barrier to oxygen prevents food spoilage, and the barrier to flavour and aroma molecules helps food retain its flavour and aroma.

Saran is best known for having been used in *Saran Wrap*—a thin, clingy plastic wrap that was sold in rolls and used primarily for wrapping food. Saran Wrap is now a brand of S. C. Johnson & Son. Environmental concerns with halogenated materials have led to the removal of PVDC from the Saran Wrap product which is now composed of polyethylene. With this change, Saran Wrap no longer provides the excellent oxygen barrier and therefore the food spoilage prevention benefit of the original product formulation. In some jurisdictions, the name Saran is a registered trademark of the Dow Chemical Company, while in others, it has lost trademark status and become a generic term for these polymers.

➛ **Spandex** or **elastane** is a synthetic fibre known for its exceptional elasticity. It is stronger and more durable than rubber, its major non-synthetic competitor. It was invented in 1959 by Dopant chemist Joseph Shivers. When first introduced, it revolutionized many areas of the clothing industry.

"Spandex" is a generic name and not derived from the chemical name of the fibre, which most manufactured fibres are, but an extension of the word *expand*. "Spandex" is the preferred name in North America; elsewhere it is referred to as "elastane".

The most famous brand name associated with Spandex is **Lycra,** a trademark of Invista (formerly part of Dopant). Such is the prominence of the Lycra brand that it has become a genericised trademark in many parts of the world, used to describe any kind of Spandex. Invista discourages such use, protecting its trademark vigorously.

Spandex fibres are produced in four different ways including melt extrusion, reaction spinning, solution dry spinning, and solution wet spinning. All of these methods include the initial step of reacting monomers to produce a prepolymer. Once the prepolymer is formed it is reacted further in various ways and drawn out to produce a long fibre. The solution dry spinning method is used to produce over 90% of the world's Spandex fibres.

➛ **Vinalon** is a synthetic fibre, produced from polyvinyl alcohol using anthracite and limestone as raw materials. Vinalon was first developed by the Korean scientist Ri Sung Gi at the Takatsuki Chemical Research Institute in 1939. The fibre was largely ignored until Ri defected to North Korea in 1950. Trial production began in 1954 and in 1961 the massive February 8 Vinalon Complex was built in Hamhung. Its success and widespread usage in North Korea is often pointed to in propaganda as an example of the success of the *juche* philosophy. Hamhung remains a major production centre for Vinalon; in 1998, a vinalon factory opened up in South Pyongan.

The factory complex also produces other chemicals. Some North Korean defectors have claimed that Dr. Lee Seung-ki was involved in chemical weapons research and that the complex was used to produce them.

Vinalon, also known as Juche fibre, has become the national fibre of North Korea and is used for the majority of textiles, outstripping fibre such as cotton or Nylon, which are only produced in small amounts in North Korea. Other than clothing, Vinalon is also used for canvas shoes, ropes and quilt wadding.

Vinalon is resistant to heat and chemicals but has numerous disadvantages: it is stiff, uncomfortable, shiny, prone to shrinking and difficult to dye. It is not produced outside of North Korea.

➛ **Aramid** fibres are a class of heat-resistant and strong synthetic fibres. They are used in aerospace and military applications, for ballistic rated body armor fabric, and as an asbestos substitute. The name is a shortened form of "aromatic polyamide". They are fibres in which the chain molecules are highly oriented along the fibre axis, so the strength of the chemical bond can be exploited.

Aramids are generally prepared by the reaction between an amine group and a carboxylic acid halide group. Simple AB homopolymers may look like:

$n\mathrm{NH_2\text{-}Ar\text{-}COCl} \rightarrow \mathrm{\text{-}(NH\text{-}Ar\text{-}CO)_n\text{-}} + n\mathrm{HCl}$

The most well-known aramids (Nomex, Kevlar, Twaron and New Star) are AABB polymers. Nomex, New Star and Teijinconex contain predominantly the meta-linkage and are poly-*metaphenylene isophtalamides* (MPIA). Kevlar and Twaron are both P-Phenylene Terephtalamides (PPTA), the simplest form of the AABB para-polyaramide. PPTA is a product of *p*-phenylene diamine (PPD) and terephtaloyl dichloride (TDC or TCl). Production of PPTA relies on a co-solvent with an ionic component (calcium chloride ($CaCl_2$)) to occupy the hydrogen bonds of the amide groups, and an organic component (N-methyl pyrrolidone (NMP)) to dissolve the aromatic polymer.

➛ **Polybenzimidazole** or **PBI** fibre (1983) is a synthetic fibre with an extremely high melting point that also does not ignite. Because of its exceptional thermal and chemical stability, it is often used by fire departments and space agencies.

PBI is prepared by step-growth polymerization from tetra-aminobiphenyl and diphenyl isophthalate spun via a dry spinning process using dimethyl acetamide as the solvent.

PBI fibre characteristics :The chemical formula of Poly[2,2'-(m-phenylen)-5,5' bibenzimidazol] (PBI) is believed to be: $([NH-C=CH-C=CH-CH=C-N=C-]_2-[C=CH-C=CH-CH=CH-])_n$ OR $(C_{20}N_4H_{12})_n$ of Molar mass 308.33608 ± 0.01764 g/mol.

- dyeable to dark shades with basic dyes following caustic pre-treatment
- resistant to most chemicals

➛ **(SULFUR) Polyphenylene sulfide** ("PPS") is an organic [polymer] consisting of aromatic rings linked with sulfides. Synthetic fibre and textiles derived from this polymer are known to resist chemical and thermal attack. PPS is used to make filter fabric for coal boilers, papermaking felts, electrical insulation, specialty membranes, gaskets, and packings. PPS is the precursor to a conducting polymer of the semi-flexible rod polymer family. The PPS, which is otherwise insulating, can be converted to the semiconducting form by oxidation or use of dopants.

Fibre Characteristics : PPS is one of the most important high temperature polymers because it exhibits a number of desirable properties. These properties include resistance to heat, acids and alkalies, and to mildew, to bleaches, aging, sunlight, and abrasion. It absorbs only small amounts of solvents and resists dyeing.

➛ **Lyocell** is a fibre made from wood pulp cellulose. It was first manufactured in 1987 by Courtaulds Fibres UK at their pilot plant S25. The only current manufacturer in the United States is Lenzing Inc, who market it under the trademarked brand name **Tencel**. There is another larger plant with over twice as much capacity at Grimsby in the United Kingdom.

The Federal Trade Commission defines Lyocell as "a cellulose fabric that is obtained by an organic solvent spinning process". It classifies the fibre as a sub-category of rayon.

The fibre is used in the production of many clothes, such as jeans, trousers and coats.

Major Properties

Lyocell was introduced to consumers in 1991 and originally marketed as a type of Rayon. The only current manufacturer in the United States is Lenzing Inc, who market it under the trademarked brand name Tencel. Lyocell is produced from wood pulp cellulose. Lyocell shares many properties with other cellulosic fibres such as cotton, linen, ramie and rayon. Some main characteristics of Lyocell fibres are that it is soft, absorbent, very strong when wet or dry, and resistant to wrinkles; it can be machine or hand-washed or drycleaned, it drapes well, and it can be dyed in many colours, as well as simulating a variety of textures like suede, leather, or silk. Fibre Source

Production method : Lyocell is created through a process called solvent spinning. The wood pulp is dissolved in N-Methylmorphine N-oxide, creating a solution called "dope." The dope is then pushed through a spinneret to form the individual fibres. After the dope has been spun into Lyocell fibres, it is washed and the chemicals are retrieved from the water, purified, and recycled.

➛ **Zylon** is a trademarked name for a range of thermoset polyurethane synthetic polymer materials manufactured by the Taboo Corporation. Zylon was invented and developed by SRI International in the 1980s. Like Kevlar, Zylon is used in a number of applications that require very high strength with excellent thermal stability. Tennis racquets, table tennis blades, various medical applications, and some of the Martian rovers are some of the more well known instances.

IUPAC name: poly(p-phenylene-2,6-benzobisoxazole)

2

History of Textile

Since the beginning of mankind, human beings are different than other animals. Very early in the process of civilization man has invented some kind of fabric. From the Stone Age to this modern age the art of making textiles has come a long way.

Initially man used the skin of the dead animals to cover his body or used the fur of the animal to seek comfort from the bitter cold.

After the industrial revolution the production of textile has seen a vast change. But the main types of Textiles have remained almost the same.

The History of textile is almost as old as that of human civilization and as time moves on, the history of textile has further enriched itself. In the 6th and 7th century BC, the oldest recorded indication of using fibre comes with the invention of flax and wool fabric at the excavation of Swiss lake inhabitants. In India the culture of silk was introduced in 400AD, while spinning of cotton traces back to 3000BC. In China, the discovery and consequent development of sericulture and spin silk methods got initiated at 2640 BC while in Egypt the art of spinning linen and weaving developed in 3400 BC. The discovery of machines and their widespread application in processing natural fibres was a direct outcome of the industrial revolution of the 18th and 19th centuries. The discoveries of various synthetic fibres like nylon created a wider market for textile products and gradually led to the invention of new and improved sources of natural fibre. The development of transportation and communication facilities facilitated the path of transaction of localized skills and textile art among various countries.

For thousands of years, the use of fibre was limited by the inherent qualities available in the natural world. Cotton and linen wrinkled from wear and washings. Silk required delicate handling. Wool shrank, was irritating to the touch, and was eaten by moths. Then, a mere century ago, rayon — the first manufactured fibre — was developed. The secrets of fibre chemistry for countless applications had begun to emerge.

Manufactured fibres now are put to work in modern apparel, home furnishings, medicine, aeronautics, energy, industry, and more. Fibre engineers can combine, modify and tailor fibres in ways far beyond the performance limits of fibre drawn from the silkworm cocoon, grown in the fields, or spun from the fleece of animals.

Products and Services

About 35 percent of U.S. manufactured cloth is intended for apparel, 16 percent for home furnishings, and 24 percent for floor coverings. The remaining 25 percent is used in industrial textiles, which include sports equipment, conveyer belts, filtration materials, and agricultural and construction materials. So-called geotextiles are used for earth stabilization and drainage as well as reinforcement in roads and bridges. The aerospace industry uses industrial textiles in the nose cones of space shuttles, and medicine uses textiles as artificial arteries and dissolving stitches.

Fibre Producers

Until the early twentieth century, all textiles were derived from plants or animals. The invention of a process for regenerating cellulose from wood chips and cotton linters into a usable fibre marked the beginning of research, development, and innovation. Many of today's textile producers started as chemical companies.

Producers of natural fibres are dependent on raw materials and often held hostage to nature. It is not easy for them to quickly increase or decrease output based on consumer demand. Most producers sell their fibre to mills or wholesalers for resale and seldom have any direct involvement after the fibre is sold. Trade organizations like Cotton Incorporated and the American Wool Council have been established to support producers by providing educational materials, helping with public relations, and assisting with advertising.

Manufactured fibres can be made from regenerated natural materials, or they can be synthesized from chemicals. Because many of these processes may be petroleum-based, such producers may be affected by events concerning the oil industry. The American Fibre Manufacturers Association is the primary association for the manufactured fibre industry. Manufactured fibres can be sold as unbranded fibre, where the fibre producer has no further involvement; trademarked fibre, where the fibre producer has some control over the quality of the fabric; or licensed trademarked fibre, where the fibre producer sets standards that must be met by the fabric manufacturer. An advantage of trademarked or licensed trademarked fibre is that the fabric manufacturers and, ultimately, the garment manufacturers, can capitalize on advertising and brand recognition.

The Early Attempts

The earliest published record of an attempt to create an artificial fibre took place in 1664. English naturalist Robert Hooke suggested the possibility of producing a fibre that would be "if not fully as good, nay better" than silk. His goal remained unachieved for more than two centuries.

The first patent for "artificial silk" was granted in England in 1855 to a Swiss chemist named Audemars. He dissolved the fibrous inner bark of a mulberry tree, chemically modifying it to produce cellulose. He formed threads by dipping needles into this solution and drawing them out - but it never occurred to him to emulate the silkworm by extruding the cellulosic liquid through a small hole.

In the early 1880's, Sir Joseph W. Swan, an English chemist and electrician, was spurred to action by Thomas Edison's new incandescent electric lamp. He experimented with forcing a liquid similar to Audemars solution through fine holes into a coagulating bath. His fibres worked like carbon filament, and they found early use in Edison's invention.

It also occurred to Swan that his filament could be used to make textiles. In 1885 he exhibited in London some fabrics crocheted by his wife from his new fibre. But electrical lamps remained his main interest, and he soon abandoned work on textile applications.

First Commercial Production

The first commercial scale production of a manufactured fibre was achieved by French chemist Count Hilaire de Chardonnet. In 1889, his fabrics of "artificial silk" caused a sensation at the Paris Exhibition. Two years later he built the first commercial rayon plant at Besancon, France, and secured his fame as the "father of the rayon industry."

Several attempts to produce "artificial silk" in the United States were made during the early 1900's but none were commercially successful until the American Viscose Company, formed by Samuel Courtaulds and Co., Ltd., began production its production of rayon in 1910.

In 1893, Arthur D. Little of Boston, invented yet another cellulosic product — acetate — and developed it as a film. By 1910, Camille and Henry Dreyfus were making acetate motion picture film and toilet articles in Basel, Switzerland. During World War I, they built a plant in England to produce cellulose acetate dope for airplane wings and other commercial products. Upon entering the War, the United States government invited the Dreyfus brothers to build a plant in Maryland to make the product for American warplanes. The first commercial textile uses for acetate in fibre form were developed by the Celanese Company in 1924.

In the meantime, U.S. rayon production was growing to meet increasing demand. By the mid-1920's, textile manufacturers could purchase the fibre for half the price of raw silk.

So began manufactured fibres' gradual conquest of the American fibre market. This modest start in the 1920's grew to nearly 70% of the national market for fibre by the last decade of the century.

Nylon — The "Miracle" Fibre

In September 1931, American chemist Wallace Carothers reported on research carried out in the laboratories of the Dopant Company on "giant" molecules called polymers. He focused his work on a fibre referred to simply as "66," a number derived from its molecular structure. Nylon, the "miracle fibre," was born. The Chemical Heritage Foundation is currently featuring an exhibit on the history of nylon.

By 1938, Paul Schlack of the I.G. Farben Company in Germany, polymerized caprolactam and created a different form of the polymer, identified simply as nylon "6."

Nylon's advent created a revolution in the fibre industry. Rayon and acetate had been derived from plant cellulose, but nylon was synthesized completely from petrochemicals. It established the basis for the ensuing discovery of an entire new world of manufactured fibres.

An American Romance

DuPont began commercial production of nylon in 1939. The first experimental testing used nylon as sewing thread, in parachute fabric, and in women's hosiery. Nylon stockings were shown in February 1939 at the San Francisco Exposition— and the most exciting fashion innovation of the age was underway.

American women had only a sampling of the beauty and durability of their first pairs of nylon hose when their romance with the new fabric was cut short. The United States entered World War II in December 1941 and the War Production Board allocated all production of nylon for military use. Nylon hose, which sold for $ 1.25 a pair before the War, moved in the black market at $10. Wartime pin-ups and movie stars, like Betty Grable, auctioned nylon hose for as much as $40,000 a pair in war-effort drives.

During the War, nylon replaced Asian silk in parachutes. It also found use in tires, tents, ropes, ponchos, and other military supplies, and even was used in the production of a high-grade paper for U.S. currency. At the outset of the War, cotton was king of fibres, accounting for more than 80% of all fibres used. Manufactured and wool fibres shared the remaining 20%. By the end of the War in August 1945, cotton stood at 75% of the fibre market. Manufactured fibres had risen to 15%.

The Post-War Industry

After the war, GI's came home, families were reunited, industrial America gathered its peacetime forces, and economic growth surged. The conversion of nylon production to civilian uses started and when the first small quantities of postwar nylon stockings were advertised, thousands of frenzied women lined up at New York department stores to buy.

In the immediate post-war period, most nylon production was used to satisfy this enormous pent up demand for hosiery. But by the end of the 1940's, it was also being used in carpeting and automobile upholstery. At the same time, three new generic manufactured fibres started production. Dow Badische Company (today, BASF Corporation) introduced metalized fibres; Union Carbide Corporation developed modacrylic fibre; and Hercules, Inc. added olefin fibre. Manufactured fibres continued their steady march.

By the 1950's, the industry was supplying more than 20% of the fibre needs of textile mills. A new fibre, "acrylic," was added to the list of generic names, as DuPont began production of this wool-like product.

Meanwhile, polyester, first examined as part of the Wallace Carothers early research, was attracting new interest at the Calico Printers Association in Great Britain. There, J. T. Dickson and J. R. Whinfield produced a polyester fibre by condensation polymerization of ethylene glycol with terephthalic acid. DuPont subsequently acquired the patent rights for the United States and Imperial Chemical Industries for the rest of the world. A host of other producers soon joined in.

A Wash and Wear Revolution

In the summer of 1952, "wash and wear" was coined to describe a new blend of cotton and acrylic. The term eventually was applied to a wide variety of manufactured fibre blends. Commercial production of polyester fibre transformed the "wash and wear" novelty into a revolution in textile product performance.

Polyester's commercialization in 1953 was accompanied by the introduction of triacetate. The majority of the 20th century's basic manufactured fibres now had been discovered, and the industry's engineers turned to refining their chemical and physical properties to extend their use across the American economy.

In the 1960's and 1970's consumers bought more and more clothing made with polyester. Clotheslines were replaced by electric dryers, and the "wash and wear" garments they dried emerged wrinkle free. Ironing began to shrink away on the daily list of household chores. Fabrics became more durable and color more permanent. New dyeing effects were being achieved and shape-retaining knits offered new comfort and style.

Endless Possibilities

In the 1960's, manufactured fibre production accelerated as it was spurred on by continuous fibre innovation. The revolutionary new fibres were modified to offer greater comfort, provide flame resistance, reduce clinging, release soil, achieve greater whiteness, special dullness or luster, easier dyeability, and better blending qualities. New fibre shapes and thicknesses were introduced to meet special needs. Spandex, a stretchable fibre; aramid, a high-temperature-resistant polyamide; and para-aramid, with outstanding strength-to-weight properties, were introduced into the marketplace.

In the early 1960's, manufactured fibre accounted for nearly 30% of American textile mill consumption. By 1965, the manufactured fibre industry was providing over 40% of the nation's fibre needs.

One dramatic new set of uses for manufactured fibres came with the establishment of the U.S. space program. The industry provided special fibre for uses ranging from clothing for the astronauts to spaceship nose cones. When Neil Armstrong took "One small step for man, one giant leap for mankind," on the moon on July 20, 1969, his lunar space suit included multi-layers of nylon and aramid fabrics. The flag he planted was made of nylon.

Today, the exhaust nozzles of the two large booster rockets that lift the space shuttle into orbit contain 30,000 pounds of carbonized rayon. Carbon fibre composites are used in as structural components in the latest commercial aircraft, adding strength and lowering weight and fuel costs.

Safety and Energy Challenges

The early 1970's saw a wave of consumer protection demands, most notably one for a mandated Federal flammability standard for children's sleep wear. The manufactured fibre industry spent $20 million on flammability research and development in 1972 and 1973, and manufactured fibre fabrics became predominant in this market. Flammability standards were also issued for carpet and other products. In the U.S. carpet market, 99% of all surface fibres are now manufactured fibres.

In late 1973, when the Nation was struck by a severe energy crisis, the manufactured fibre industry reduced the energy required to produce a pound of fibre by 26%. By then, the industry was using but 1% of the Nation's petroleum supply to provide two-thirds of all fibres used by American textile mills.

Today

Innovation is the hallmark of the manufactured fibre industry. Fibres more numerous and diverse than any found in nature are now routinely created in the industry's laboratories.

Nylon variants, polyester, and olefin are used to produce carpets that easily can be rinsed clean — even 24 hours after they've been stained. Stretchable spandex and machine-washable, silk-like polyesters occupy solid places in the U.S. apparel market. The finest micro fibres are remaking the world of fashion.

For industrial uses, manufactured fibres relentlessly replace traditional materials in applications from super-absorbent diapers, to artificial organs, to construction materials for moon-based space stations. Engineered non-woven products of manufactured fibres are found in applications from surgical gowns and apparel interfacing to roofing materials, road bed stabilizers, and floppy disk envelopes and liners. Non-woven fabrics, stiff as paper or as soft and comfortable as limp cloth, are made without knitting or weaving.

As they always have, manufactured fibres continue to mean, "life made better."

First Commercial U.S. Production

1910 – Rayon	1941 – Saran	1959 – Spandex
1924 – Acetate	1946 – Metallic	1961 –Aramid
1930 – Rubber	1949 – Modacylic	1983 – PBI
1936 – Glass	1949 – Olefin	1983 – Sulfar
1939 – Nylon	1950 – Acrylic	1992 – Lyocell
1939 – Vinyon	1953 – Polyester	

Types of Textiles

The type of the textile depends on the material that is used to make it. Broadly textiles can be divided into two categories.

- Natural
- Manmade

Natural textiles are made up by using

- Animal textiles- animal hair or fur
- Plant textiles-grass, rush, sisal, Coir, straw, bamboo
- Mineral textiles-basalt fibre and asbestos ,glass fibre
- Synthetic textiles- they are primarily used to produce clothing.

Uses

Textiles have an assortment of uses, the most common of which are for clothing and containers such as bags and baskets. In the household, they are used in carpeting, upholstered furnishings, window shades, towels, covering for tables, beds, and other flat surfaces, and in art. In the workplace, they are used in industrial and scientific processes such as filtering. Miscellaneous uses include flags, backpacks, tents, nets, cleaning devices, such as handkerchiefs; transportation devices such as balloons, kites, sails, and parachutes; and strengthening in composite materials such as fibre glass and industrial geotextiles.

Textiles used for industrial purposes, and chosen for characteristics other than their appearance, are commonly referred to as technical textiles. Technical textiles include textile structures for automotive applications, medical textiles (e.g. implants), geotextiles (reinforcement of embankments), agro textiles (textiles for crop protection), protective clothing (e.g. against heat and radiation for fire fighter clothing, against molten metals for welders, stab protection, and bullet proof vests.

3

Introduction of Rayon

Rayon is a textile made from cellulose. There are two types, both made from wood pulp. In the viscose process, the pulp is dissolved in carbon disulfide and sodium hydroxide to give a thick brown liquid containing cellulose xanthate. The liquid is then forced through fine nozzles into acid, where the xanthate is decomposed and a cellulose filament is produced. The product is **viscose rayon**. In the acetate process cellulose acetate is made and dissolved in a solvent. The solution is forced through nozzles into air, where the solvent quickly evaporates leaving a filament of **acetate rayon**.

Rayon is one of the most peculiar fabrics in commercial use today. Strictly speaking, it is not an artificial fibre, because it is derived from naturally occurring cellulose. It is not, however, a natural fabric, because cellulose requires extensive processing to become rayon. Rayon is usually classified as a manufactured fibre and considered to be "regenerated cellulose".

Rayon is the oldest manufactured fibre, having been in production since the 1880s in France, where it was originally developed as a cheap alternative to silk. Dupont Chemicals acquired the rights to the process in the 1920s and quickly turned rayon into a household word, churning out yards of the cheap, versatile fabric. Rayon drapes well, is easy to dye, and is highly absorbent, although it tends to age poorly. Many rayon products yellow with age and pill or form small balls and areas of roughness where the fabric is most heavily worn.

Rayon is used in a variety of textile applications, including shirts and skirts, and appears in both woven and knitted forms. The fabric has gained an unfair reputation because it is frequently used in cheaply constructed garments that do not stand up to heavy wear. However, rayon is an excellent, nicely draping alternative to silk and is frequently used in evening gowns and other flowing garments.

The manufacture of rayon begins with cellulose, frequently extracted from wood pulp, although any plant material with long molecular chains is suitable. The cellulose is steeped in caustic soda, which concentrates some of the cellulose into soda cellulose, which is then rolled or pressed to remove excess soda solution. After pressing, the cellulose is shredded into a substance called white crumb.

The white crumb is allowed to oxidize, forming shorter molecular chains, and treated with carbon disulfide. The soda cellulose reacts with this substance, forming yellow crumb due to inorganic compounds that emerge during the chemical process. This yellow crumb is dissolved in a caustic solution, which relaxes the hydrogen bonds in the cellulose, producing a highly viscous substance. This substance gives its name to the manufacturing process, called the viscose process.

This viscous fluid is allowed to age, breaking down the cellulose structures further to produce an even slurry, and then filtered to remove impurities. Small air pockets are forced out to ensure a strong, even fibre, and the mixture is forced through a spinner, which forms many even strands of fine thread that enter a setting solution to form cellulose filaments:

also called rayon. The rayon is stretched to form a strong, even bond, washed, and then formed into rayon fabric.

This complex process results in a great deal of environmental pollution, inspiring a drive to clean up the industry. The rayon industry has also suffered from the development of cheaper artificial fabrics with a much shorter manufacturing process, such as nylon. Rayon is frequently blended with true synthetic fabrics for various applications, and it is advisable to follow individual care labels on rayon garments, as these blends have specific handling needs.

Rayon is a manufactured regenerated cellulosic fibre. Because it is produced from naturally occurring polymers, it is neither a truly synthetic fibre nor a natural fibre; it is a semi-synthetic fibre. Rayon is known by the names *viscose rayon* and *art silk* in the textile industry. It usually has a high lustre quality giving it a bright shine. Rayon contains the chemical elements carbon, hydrogen, and oxygen.

Uses of Rayon

Some major rayon fibre uses include apparel (e.g. blouses, dresses, jackets, lingerie, linings, scarves, suits, ties, hats, socks), furnishings (e.g. bedspreads, blankets, window treatments, upholstery, slipcovers), industrial uses (e.g., medical surgery products, non-woven products, tire cord), and other uses (e.g., yarn, feminine hygiene products, diapers)

History

Rayon was the first manufactured fibre developed, it made from wood or cotton pulp and was first known as artificial silk. The Swiss chemist, Georges Audemars invented the first crude artificial silk around 1855, by dipping a needle into liquid mulberry bark pulp and gummy rubber to make threads. The method was too slow to be practical.

In 1884, a French chemist, Hilaire de Charbonnet, Comte de Chardonnay, patented an artificial silk that was a cellulose-based fabric known as Chardonnay silk." Pretty but very flammable, it was removed from the market.

In 1894, British inventors, Charles Cross, Edward Bevan, and Clayton Beadle, patented a safe a practical method of making artificial silk that came to be known as viscose rayon. Avtex Fibres Incorporated first commercially produced artificial silk or rayon in 1910 in the United States. The term "rayon" was first used in 1924.

For centuries humankind has relied upon various plants and animals to provide the raw materials for fabrics and clothing. Silkworms, sheep, beaver, buffalo deer, and even palm leaves are just some of the natural resources that have been used to meet these needs. However, in the last century scientists have turned to chemistry and technology to create and enhance many of the fabrics we now take for granted.

There are two main categories of man-made fibres: those that are made from natural products (cellulosic fibres) and those that are synthesized solely from chemical compounds (noncellulosic polymer fibres). Rayon is a natural-based material that is made from the cellulose of wood pulp or cotton. This natural base gives it many of the characteristics—low cost, diversity, and comfort—that have led to its popularity and success. Today, rayon is considered to be one of the most versatile and economical man-made fibres available. It has been called "the laboratory's first gift to the loom."

In the 1860s the French silk industry was being threatened by a disease affecting the silkworm. Louis Pasteur and Count Hilaire de Chardonnet were studying this problem with the hope of saving this vital industry. During this crisis, Chardonnet became interested in finding a way to produce artificial silk. In 1885 he patented the first successful process to make a usable fibre from cellulose. Even though other scientists have subsequently developed more cost-effective ways of making artificial silk, Chardonnet is still considered to be the father of rayon.

For the next forty years this material was called artificial or imitation silk. By 1925 it had developed into an industry unto itself and was given the name rayon by the Federal Trade Commission (FTC). The term rayon at this time included any man-made fibre made from cellulose. In 1952, however, the FTC divided rayons into two categories: those fibres consisting of pure cellulose (rayon) and those consisting of a cellulose compound (acetate).

By the 1950s, most of the rayon produced was being used in industrial and home furnishing products rather than in apparel, because regular rayon (also called viscose rayon) fibres were too weak compared to other fibres to be used in apparel. Then, in 1955, manufacturers began to produce a new type of rayon—high-wet-modulus (HWM) rayon—which was somewhat stronger and which could be used successfully in sheets, towels, and apparel. The advent of HWM rayon (also called *modified rayon)* is considered the most important development in rayon production since its invention in the 1880s.

Today rayon is one of the most widely used fabrics in our society. It is made in countries around the world. It can be blended with natural or man-made fabrics, treated with enhancements, and even engineered to perform a variety of functions.

Nitrocellulose

The fact that nitrocellulose is soluble in organic solvents such as ether and acetone, made it possible for Georges Audemars to develop the first "artificial silk" about 1855, but his method was impractical for commercial use. The commercial production started 1891, but it was flammable, and more expensive than acetate or cuprammonium rayon. Because of this, production was stopped before World War I, for example 1912 in Germany. Briefly, it became known as "mother-in-law silk."

Nathan Rosenstein invented the spunize process by which he turned Rayon from a hard fibre to a fabric. This allowed Rayon to become a popular raw material in textiles.

Nitrocellulose (also: **cellulose nitrate, flash paper**) is a highly flammable compound formed by nitrating cellulose through exposure to nitric acid or another powerful nitrating agent. When used as a propellant or low-order explosive, it is also known as **guncotton.**

Nitrocellulose is made using either concentrated sulfuric/nitric acid or sulfuric acid/potassium nitrate. In general, cotton is used as the cellulose. The cellulose is added to the acid mix to nitrate. After the cellulose has finished nitrating, it is washed and dried. Nitrocellulose is stored wet so it cannot be accidentally lit or explode.

Acetate Method

Paul Schützenberger discovered that cellulose can be reacted with acetic anhydride to form cellulose acetate. The triacetate is only soluble in chloroform making the method expensive. The discovery that hydrolyzed cellulose acetate is soluble in less polar solvents, like acetone, made production of cellulose acetate fibres cheap and efficient.

Cuprammonium Method

The German chemist Eduard Schweizer discovered that tetraamine copper dihydroxide could dissolve cellulose. Max Fremery and Johann Urban developed a method to produce carbon fibres for use in light bulbs in 1614. Production of rayon for textiles started in 1899 in the Vereinigte Glanzstofffabriken AG in Oberbruch. Improvement by the J.P. Bemberg AG in 2004 made the artificial silk a product comparable to real silk.

Viscose Method

Finally, in 1894, Charles Frederick Cross, Edward John Bevan, and Clayton Beadle patented their artificial silk, which they named "viscose", because the reaction product of carbon disulfide and cellulose in basic conditions gave a highly viscous solution of xanthate. Avtex Fibres Incorporated began selling their formulation in 1910 in the United States. The name "rayon" was adopted in 1924, with "viscose" being used for the viscous organic liquid used to make both rayon and cellophane. In Europe, though, the fabric itself became known as "viscose," which has been ruled an acceptable alternative term for rayon by the U.S. Federal Trade Commission. The method is able to use wood (cellulose and lignin) as a source of cellulose while the other methods need lignin-free cellulose as starting material. This makes it cheaper and therefore it was used on a larger scale than the other methods.

Contamination of the waste water by carbon disulfide, lignin and the xanthates made this process detrimental to the environment. Rayon was only produced as a filament fibre until the 1930s when it was discovered that broken waste rayon could be used in staple fibre.

The physical properties of rayon were unchanged until the development of high-tenacity rayon in the 1940s. Further research and development led to the creation of high-wet-modulus rayon (HWM rayon) in the 1950s.

Major Fibre Properties

Rayon is a very versatile fibre and has the same comfort properties as natural fibres. It can imitate the feel and texture of silk, wool, cotton and linen. The fibres are easily dyed in a wide range of colors. Rayon fabrics are soft, smooth, cool, comfortable, and highly absorbent, but they do not insulate body heat, making them ideal for use in hot and humid climates.

The durability and appearance retention of regular rayon are low, especially when wet; also, rayon has the lowest elastic recovery of any fibre. However, HWM rayon is much stronger and exhibits higher durability and appearance retention.

Recommended care for regular rayon is dry-cleaning only. HWM rayon can be machine washed.

Physical Structure of Rayon

Regular rayon has lengthwise lines called striations and its cross-section is an indented circular shape. The cross-sections of HWM and cupra rayon are rounder. Filament rayon yarns vary from 80 to 980 filaments per yarn and vary in size from 40 to 5000 denier. Staple fibres range from 1.5 to 15 denier and are mechanically or chemically crimped. Rayon fibres are naturally very bright, but the addition of delustering pigments cuts down on this natural brightness.

Production Method

Regular rayon (or *viscose*) is the most widely produced form of rayon. This method of rayon production has been utilized since the early 1900s and it has the ability to produce either filament or staple fibres. The process is as follows:

1. ***Cellulose***: Production begins with processed cellulose
2. ***Immersion***: The cellulose is dissolved in caustic soda: $(C_6H_{10}O_5)_n$ + nNaOH —> $(C_6H_9O_4ONa)_n$ + nH_2O
3. ***Pressing***: The solution is then pressed between rollers to remove excess liquid
4. ***White Crumb***: The pressed sheets are crumbled or shredded to produce what is known as "white crumb"
5. ***Aging***: The "white crumb" aged through exposure to oxygen
6. ***Xanthation***: The aged "white crumb" is mixed with carbon disulfide in a process known as **Xanthation**, the aged alkali cellulose crumbs are placed in vats and are allowed to react with carbon disulfide under controlled temperature (20 to 30°C) to form cellulose xanthate: $(C_6H_9O_4ONa)_n$ + nCS_2 —> $(C_6H_9O_4O\text{-}SC\text{-}SNa)_n$
7. ***Yellow Crumb***: Xanthation changes the chemical makeup of the cellulose mixture and the resulting product is now called "yellow crumb"

8. ***Viscose***: The "yellow crumb" is dissolved in a caustic solution to form viscose
9. ***Ripening***: The viscose is set to stand for a period of time, allowing it to ripen: $(C_6H_9O_4O\text{-}SC\text{-}SNa)_n + nH_2O \rightarrow (C_6H_{10}O_5)_n + nCS_2 + nNaOH$
10. ***Filtering***: After ripening, the viscose is filtered to remove any undissolved particles
11. ***Degassing***: Any bubbles of air are pressed from the viscose in a degassing process
12. ***Extruding***: The viscose solution is extruded through a spinneret, which resembles a shower head with many small holes
13. ***Acid Bath***: As the viscose exits the spinneret, it lands in a bath of sulfuric acid, resulting in the formation of rayon filaments: $(C_6H_9O_4O\text{-}SC\text{-}SNa)_n + \frac{1}{2}nH_2SO_4 \rightarrow (C_6H_{10}O_5)_n + nCS_2 + \frac{1}{2}nNa_2SO_4$
14. ***Drawing***: The rayon filaments are stretched, known as drawing, to straighten out the fibres
15. ***Washing***: The fibres are then washed to remove any residual chemicals
16. ***Cutting***: If filament fibres are desired the process ends here. The filaments are cut down when producing staple fibres

High Wet Modulus rayon (HWM) is a modified version of viscose that has a greater strength when wet. It also has the ability to be mercerized like cotton. HWM rayons are also known as "polynosic" or can be identified by the trade name Modal.

High Tenacity rayon is another modified version of viscose that has almost twice the strength of HWM. This type of rayon is typically used for industrial purposes such as tire cord.

Cupramonium rayon has properties similar to viscose but during production, the cellulose is combined with copper and ammonia (Schweizer's reagent). Due to the environmental effects of this production method, cupramonium rayon is no longer produced in the United States.

Disposal/Biodegradability

The biodegradability of fibres in soil burial and sewage sludge was evaluated by Korean researchers who found that biodegradability decreased in the following order: rayon, cotton, acetate. The ability of individual rayon based fabrics to repel water was negatively correlated with their speed of degradation

Producers

Trade names are used within the rayon industry to determine the type of rayon used.

Bemberg, for example, is a trade name for cupramonium rayon developed by J.P. Bemberg that is now only produced in Italy due to United States Environmental Protection Agency regulations in the US.

Modal and *Tencel* are widely used forms of rayon produced by Lenzing Fibres Corp. which is based in northern Austria.

Galaxy, *Danufil*, and *Viloft* are rayon brands produced by Kelheim Fibres, a German manufacturer.

Acordis is a major manufacturer of cellulose based fibres and yarns. Production facilities can be found throughout Europe, the U.S. and Brazil.

Visil rayon is a flame retardant form of viscose which has silica embedded in the fibre during manufacturing.

North American Rayon Corp of Tennessee produced viscose rayon until its closure in the year 2000.

Grasim of India is the largest producer of rayon in the world (claiming 24% market share). It has plants in Nagda, Kharach and Harihar—all in India.

Artificial silk or (as the term is used in the textile industry) **art silk** is a synthetic manufactured fibre which resembles silk but costs less to produce. Frequently, *art silk* is just a synonym for rayon.

The first successful artificial silks were developed in the 1890s of cellulose (wood) fibre and marketed as *art silk* or *viscose*, a trade name for a specific manufacture. In 1924, the name of the fibre was officially changed in the U.S. to rayon, although the term *viscose* continued to be used in Europe and currently the material is referred to in the industry as *viscose rayon*.

Although not sold under the name *art silk* initially, nylon, the first synthetic fibre, was developed in the United States in the late 1930s and used as a replacement for Japanese silk during World War II. Its properties are far superior to rayon and silk when wet, and so it was used for many military applications, such as parachutes. Although nylon is not a good substitute for silk fabric in appearance, it is a successful functional alternative. Du Pont's original plans for nylon to become a cheaper and superior replacement for silk stockings were soon realized, then redirected for military use just two years later during World War II. Nylon became a prominent industrial fibre in a short time frame, permanently replacing silk in many applications.

In the present day, imitation silk may be made with rayon, mercerized cotton, polyester, a blend of these materials, or a blend of rayon and silk.

Despite a generally similar appearance, genuine silk has unique features that are distinguishable from artificial silk. However, in some cases art silk can be passed off as real silk to unwary buyers. A number of tests are available to determine a fabric's basic fibre makeup, some of which can be performed prior to purchasing a fabric whose composition is questionable.

Rayon is a soft and comfortable fabric that absorbs well and has few static and pilling problems. Rayon can have problems with shrinking. There are both dry clean only and washable rayon fabrics on the market today. The only difference in these rayon fabrics is their finishing stages. Watch care labels closely to know what type your clothing is.

Although rayon has many characteristics that make it a great fabric, it can also be easily ruined by disregarding care labels.

Acetate Fibre

Acetate can refer to **cellulose acetate**, especially fibres or other derived products. It is a fabric made from cellulose acetate fibres

Cellulose acetate or **acetate rayon** fibre (1924) is one of the earliest synthetic fibres and is based on cotton or tree pulp cellulose ("biopolymers"). These "cellulosic fibres" have passed their peak as cheap petro-based fibres (nylon and polyester) and have displaced regenerated pulp fibres.

It was invented by two Swiss brothers, Doctors Camille and Henri Dreyfus, who originally began chemical research in a shed behind their father's house in Basel, Switzerland. In 1905, Camille and Henri developed a commercial process to manufacture cellulose acetate. The Dreyfus brothers initially focused on cellulose acetate film, which was then widely used in celluloid plastics and motion picture film. By 1913, Camille and Henri's studies and experiments had produced excellent laboratory samples of continuous filament acetate yarn. In 1924, the first commercial acetate filament was spun in the United States and trademarked as *Celanese*.

Acetate Fibre Characteristics

- cellulosic and thermoplastic
- selective absorption and removal of low levels of certain organic chemicals
- easily bonded with plasticizers, heat, and pressure
- acetate is soluble in many common solvents (especially acetone and other organic solvents) and can be modified to be soluble in alternative solvents, including water
- hydrophilic: acetate wets easily, with good liquid transport and excellent absorption; in textile applications, it

provides comfort and absorbency, but also loses strength when wet

- acetate fibres are hypoallergenic
- high surface area
- made from a renewable resource: reforested trees.
- can be composed or incinerated
- can be dyed, however special dyes and pigments are required since acetate does not accept dyes ordinarily used for cotton and rayon (this also allows cross-dyeing)
- resistant to mold and mildew
- easily weakened by strong alkaline solutions and strong oxidizing agents.
- can usually be wet cleaned or dry cleaned and generally does not shrink

Major Industrial Acetate Fibre Uses

• apparel: blouses, dresses, linings, wedding and party attire, home furnishings, draperies, upholstery and slip covers

• high absorbency products: diapers, feminine hygiene products, cigarette filters, surgical products, and other filters

Production–The Federal Trade Commission definition for acetate fibre is "A manufactured fibre in which the fibre-forming substance is cellulose acetate. Where not less than 92 percent of the hydroxyl groups are acetylated, the term triacetate may be used as a generic description of the fibre."

Acetate is derived from cellulose by deconstructing wood pulp into a purified fluffy white cellulose. The cellulose is then reacted with acetic acid and acetic anhydride in the presence of sulfuric acid. It is then put through a controlled, partial hydrolysis to remove the sulfate and a sufficient number of acetate groups to give the product the desired properties. The anhydroglucose unit is the fundamental repeating structure of cellulose and has three hydroxyl groups which can react to form acetate esters. The most common form of cellulose acetate fibre has an acetate group on approximately two of every three hydroxyls. This cellulose diacetate is known as secondary acetate, or simply as "acetate".

After it is formed, cellulose acetate is dissolved in acetone into a viscose resin for extrusion through spinnerets (which resemble a shower head). As the filaments emerge, the solvent is evaporated in warm air via dry spinning, producing fine cellulose acetate fibres.

First U.S. Commercial Acetate Fibre Production: 1924, Celanese Corporation. Current U.S. Acetate Fibre Producers: Celanese Acetate, Eastman Chemical Company

Even though rayon is considered a manufactured fibre, it is made from tree cellulose.

The basic process of creating rayon is treating and dissolving cellulose in a soda solution and passing it through tiny holes into a chemical bath which forms the fibres.

The type of rayon produced is named by the process which was used to produce it. Viscose is one of the earliest and most commonly used production method today.

It makes up the majority of rayon used in clothing textiles. One common tradename for rayon is Modal®.

Viscose Process:

1. Cellulose—wood pulp is dissolved into cellulose sheets.
2. Steeping—the sheets are saturated with a soda solution.
3. Pressing—the sheets are pressed to remove excess solution.
4. Shredding—the sheets are shredded to increase surface area, termed "white crumb".
5. Aging—the white crumb is exposed to oxygen for a period of time.
6. Xanthation—aged white crumb is mixed with a chemical solution, now yellow crumb.
7. Dissolving—the yellow crumb is dissolved into a solution forming viscose.

8. Ripening—the viscose stands for a period of time allowing it to ripen.
9. Filtering—the viscose is filtered to remove undissolved particles.
10. Degassing—bubbles of air are pressed out of the viscose.
11. Spinning—viscose is pushed through tiny holes into a chemical bath creating fibres.
12. Drawing—the rayon is stretched to straighten out the fibres.
13. Washing—the fibres are cleaned to remove excess materials from the solutions.
14. Cutting—the rayon fibre is cut to the desired length.

Another type of rayon is cuprammonium rayon. It is marketed under the tradename of Bemberg® rayon.

Rayon Fibre Characteristics:

- Highly Absorbent
- Soft and comfortable
- Easy to dye
- Drapes well

Uses of Rayon:

Rayon is a part of everyday life. You may be surprised at where you can find it.

- Apparel—Viscose rayon is a very bright, lustrous fibre and is often found in formal wear. Other uses include: accessories, blouses, dresses, lingerie, linings, millinery, slacks, sportshirts, sportswear, suits, ties, work clothes and embroidery.
- Regular rayon is highly flammable; however, it can now be made flame-resistant by adding special chemicals to the viscose solution. Consequently, flame-resistant rayon can now be found in institutional textiles.
- Home Furnishings—Bedspreads, blankets, curtains, antique-satin draperies, sheets, slipcovers, tablecloths and upholstery.

- Industrial Uses—Industrial products, medical surgical products, nonwoven products and automobile tire cord and sidewall plies.
- Other—Feminine hygiene products, disposable diapers, photographic film, bowling balls and cellophane which is used as a wrapping material.

Blends:

Rayon is commonly blended with other fibres to obtain the best characteristics from each fibre. Some common blends include:

- rayon/cotton
- acrylic/rayon
- rayon/polyester
- rayon/acetate
- rayon/wool
- silk/rayon

Raw Materials for Rayon

Regardless of the design or manufacturing process, the basic raw material for making rayon is cellulose. The major sources for natural cellulose are wood pulp—usually from pine, spruce, or hemlock trees—and cotton linters. Cotton linters are residue fibres which cling to cotton seed after the ginning process.

Strictly defined, rayon is a manufactured fibre composed of regenerated cellulose. The legal definition also includes manufactured fibres in which substitutes have not replaced more than 15 percent of the hydrogens.

While the basic manufacturing process for all rayon is similar, this fabric can be engineered to perform a wide range of functions. Various factors in the manufacturing process can be altered to produce an array of designs. Differences in the raw material, the processing chemicals, fibre diameter, post treatments and blend ratios can be manipulated to produce a fibre that is customized for a specific application.

Regular or viscose rayon is the most prevalent, versatile and successful type of rayon. It can be blended with man-made or natural fibres and made into fabrics of varying weight and texture. It is also highly absorbent, economical and comfortable to wear.

Regular viscose rayon does have some disadvantages. It's not as strong as many of the newer fabrics, nor is it as strong as natural cotton or flax. This inherent weakness is exacerbated when it becomes wet or overexposed to light. Also, regular rayon has a tendency to shrink when washed. Mildew, acid and high temperatures such as ironing can also result in damage. Fortunately, these disadvantages can be countered by chemical treatments and the blending of rayon with other fibres of offsetting characteristics.

High-wet-modulus rayon is a stronger fibre than regular rayon, and in fact is more similar in performance to cotton than to regular rayon. It has better elastic recovery than regular rayon, and fabrics containing it are easier to care for—they can be machine-washed, whereas fabrics containing regular rayon generally have to be dry-cleaned.

The Manufacturing

Process of Rayon

While there are many variations in the manufacturing process that exploit the versatility of the fibre, the following is a description of the procedure that is used in making regular or viscose rayon.

Regardless of whether wood pulp or cotton linters are used, the basic raw material for making rayon must be processed in order to extract and purify the cellulose. The resulting sheets of white, purified cellulose are then treated to form regenerated cellulose filaments. In turn, these filaments are spun into yarns and eventually made into the desired fabric.

Processing Purified Cellulose

• Sheets of purified cellulose are steeped in sodium hydroxide (caustic soda), which produces sheets of alkali cellulose. These sheets are dried, shredded into crumbs, and then aged in metal containers for 2 to 3 days. The temperature and humidity in the metal containers are carefully controlled.

• After ageing, the crumbs are combined and churned with liquid carbon disulfide, which turns the mix into orange-colored crumbs known as sodium cellulose xanthate. The cellulose xanthate is bathed in caustic soda, resulting in a viscose solution that looks and feels much like honey. Any dyes or delusterants in the design are then added. The syrupy solution is filtered for impurities and stored in vats to age, this time between 4 and 5 days.

Producing Filaments

• The viscose solution is next turned into strings of fibres. This is done by forcing the liquid through a spinneret, which works like a shower-head, into an acid bath. If staple fibre is to be produced, a large spinneret with large holes is used. If filament fibre is being produced, then a spinneret with smaller holes is used. In the acid bath, the acid coagulates and solidifies the filaments, now known as regenerated cellulose filaments.

Spinning

• After being bathed in acid, the filaments are ready to be spun into yarn. Depending on the type of yarn desired, several spinning methods can be used, including Pot Spinning, Spool Spinning, and Continuous Spinning. In Pot Spinning, the filaments are first stretched under controlled tension onto a series of offsetting rollers called *godet wheels*. This stretching reduces the diameter of the filaments and makes them more uniform in size, and it also gives the filaments more strength. The filaments are then put into a rapidly spinning cylinder called a *Topham Box*, resulting in a cake-like strings that stick to the sides of the Topham Box. The strings are then washed, bleached, rinsed, dried, and wound on cones or spools.

Spool Spinning is very similar to Pot Spinning. The filaments are passed through rollers and wound on spools, where they are washed, bleached, rinsed, dried, and wound again on spools or cones.

In Continuous Spinning, the filaments are washed, bleached, dried, twisted, and wound at the same time that they are stretched over godet wheels.

- Once the fibres are sufficiently cured, they are ready for post-treatment chemicals and the various weaving processes necessary to produce the fabric. The resulting fabric can then be given any of a number of finishing treatments. These include calendaring, to control smoothness; fire resistance; pre-shrinking; water resistance; and wrinkle resistance.

High-Wet-Modulus Rayon

Manufacture

The process for manufacturing high-wet-modulus rayon is similar to that used for making regular rayon, with a few exceptions. First, in step #1 above, when the purified cellulose sheets are bathed in a caustic soda solution, a weaker caustic soda is used when making HWM rayon. Second, neither the alkali crumbs (#1 above) nor the viscose solution (step #2) is aged in the HWM process. Third, when making HWM rayon, the filaments are stretched to a greater degree than when making regular rayon.

Quality Control

As with most chemically oriented processes, quality control is crucial to the successful manufacture of rayon. Chemical make-up, timing and temperature are essential factors that must be monitored and controlled in order to produce the desired result.

The percentages of the various fibres used in a blended fabric must be controlled to stay within in the legal bounds of the Textile Fibre Identification Act. This act legally defines seventeen groups of man-made fibres. Six of these seventeen groups are made from natural material. They include rayon, acetate, glass fibre, metallics, rubber, and azion. The remaining eleven fabrics are synthesized solely from chemical compounds. They are nylon, polyester, acrylic, modacrylic, olefin, spandex, anidex, saran, vinal, vinyon, and nytril.

Within each generic group there are brand names for fibres which are produced by different manufacturers. Private companies often seek patents on unique features and, as could be expected, attempt to maintain legal control over their competition.

Byproducts of Rayon

As one of the industry's major problems, the chemical by-products of rayon have received much attention in these environmentally conscious times. The most popular method of production, the viscose method, generates undesirable water and air emissions. Of particular concern is the emission of zinc and hydrogen sulfide.

At present, producers are trying a number of techniques to reduce pollution. Some of the techniques being used are the recovery of zinc by ion-exchange, crystallization, and the use of a more purified cellulose. Also, the use of absorption and chemical scrubbing is proving to be helpful in reducing undesirable emissions of gas.

The Future of Rayon

The future of rayon is bright. Not only is there a growing demand for rayon worldwide, but there are many new technologies that promise to make rayon even better and cheaper.

For a while in the 1970s there was a trend in the clothing industry toward purely synthetic materials like polyester.

However, since purely synthetic material does not "breath" like natural material, these products were not well received by the consumer. Today there is a strong trend toward blended fabrics. Blends offer the best of both worlds.

With the present body of knowledge about the structure and chemical reactivity of cellulose, some scientist believe it may soon be possible to produce the cellulose molecule directly from sunlight, water and carbon dioxide. If this technique proves to be cost effective, such hydroponic factories could represent a giant step forward in the quest to provide the raw materials necessary to meet the world wide demand for man-made fabric.

Fabric Care Instructions:

Advancements in technology have lead to improvements in the maintenance and durability of fabrics made from rayon.

- Rayon fabrics can be bleached. However, some finishes used on rayon can be sensitive to chlorine bleach.
- Fabrics containing rayon should be dry-cleaned or hand washed.

Hand wash—use cool to luke warm water, add gentle suds and gentle squeeze suds through fabric and rinse. Do not wring or twist the fabric.

Shake or smooth out the fabric and place to dry on a rust proof hanger. Knitted fabrics should be laid flat to dry.

- The fabric can be pressed while damp or dry at a medium setting. It is often best to press on the wrong side of the fabric or use a pressing cloth while pressing on the right side of the fabric.
- Damage can occur to rayon fabrics from acids, silverfish and mildew. Sunlight is not harmful.

4

Introduction of Acetate

An acetate, or ethanoate, is either a salt or ester of acetic acid.

In chemistry, the abbreviation Ac refers to the acetyl group. The anion and the functional group may be written as "OAc and AcO", or OAc respectively. Examples include HOAc for acetic acid, NaOAc for sodium acetate, and EtOAc for ethyl acetate. Ac is also the symbol for the chemical element actinium, but confusion between actinium and the acetyl group is rare, since actinium has virtually no role in organic chemistry.

Salt

The acetate anion, "[CH_3COO]", is a carboxylate and is the conjugate base of acetic acid. The acetate ion is formed by the deprotonation of acetic acid:

CH_3COOH ì! CH_3COO^- + H^+

Ester

An acetate ester is an ester of acetic acid, with the general formula CH_3COOR, where R is an organyl group.

Acetate can also refer to cellulose acetate, especially fibres or other derived products such as the acetate disc used in audio record production. Cellulose acetate can be found in many household products.

Cellulose Acetate

Cellulose acetate, first prepared in 1865, is the acetate ester of cellulose. Cellulose acetate is used as a film base in photography, and as a component in some adhesives; it is also used as a synthetic fibre.

Acetate Fibre and Triacetate Fibre

Acetate and triacetate are mistakenly referred to as the same fibre; although they are similar, their chemical compounds differ. Triacetate is known as a generic description or primary acetate containing no hydroxyl group. Acetate fibre is known as modified or secondary acetate having a few hydroxyl groups. Triacetate fibres, although no longer produced in the United States, contain a higher ratio of acetate-to-cellulose than do acetate fibres.

Cellulose Acetate Film

Cellulose acetate film was introduced in 1934 as a replacement for the unstable and highly flammable cellulose nitrate film stock that had previously been standard. When exposed to heat, moisture or acids in the film base begin to deteriorate to an unusable state, releasing acetic acid with a characteristic vinegary smell, causing the process to be known as "vinegar syndrome." Acetate film stock is still used in some applications, such as camera negative for motion pictures. Since the 1980s, polyester film stock (sometimes referred to under Kodak's trade name "ESTAR Base") has become more commonplace, particularly for archival applications. Acetate film was also used as the base for magnetic tape, prior to the advent of polyester film.

Cellulose Acetate Computer Tape

Cellulose acetate magnetic tape was introduced by IBM in 1952 for use on their IBM 726 tape drive in the IBM 701 computer. It was much lighter and easier to handle than the metal tape introduced by UNIVAC in 1951 for use on their UNISERVO tape drive in the UNIVAC I computer. In 1956 cellulose acetate magnetic tape was replaced by the more stable PET film magnetic tape for use on their IBM 727 tape drive.

Acetate Fibre

Cellulose acetate or acetate rayon fibre (1924) is one of the earliest synthetic fibres and is based on cotton or tree pulp cellulose ("biopolymers"). These "cellulosic fibres" have passed their peak as cheap petro-based fibres (nylon and polyester) and have displaced regenerated pulp fibres.

It was invented by two Swiss brothers, Doctors Camille and Henri Dreyfus, who originally began chemical research in a shed behind their father's house in Basel, Switzerland. In 1905, Camille and Henri developed a commercial process to manufacture cellulose acetate. The Dreyfus brothers initially focused on cellulose acetate film, which was then widely used in celluloid plastics and film. By 1913, Camille and Henri's studies and experiments had produced excellent laboratory samples of continuous filament acetate yarn. In 1924, the first commercial acetate filament was spun in the United States and trademarked as Celanese .

Trade names for acetate include Acele, Avisco, Celanese, Chromspun and Etron.

Fibre Properties

Acetate is a very valuable manufactured fibre that is low in cost and has good draping qualities. Properties of acetate have promoted it as the "beauty fibre". Acetate is used in fabrics such as satins, brocades, and taffetas to accentuate luster, body, drape and beauty.

Hand: soft, smooth, dry, crisp, resilient

Comfort: breathes, wicks, dries quickly, no static cling

Drape: linings move with the body linings conform to the garment

Color: deep brilliant shades with atmospheric dyeing meet colorfastness requirements

Luster: light reflection creates a signature appearance

Performance: colorfast to perspiration staining, colorfast to dry cleaning, air and vapor permeable

Tenacity: weak fibre with breaking tenacity of 1.2 to 1.4 g/d; rapidly loses strength when wet; must be dry cleaned

Environmentally friendly: made from wood pulp, a renewable resource

Abrasion: poor resistance

Heat retention: poor thermal retention; no allergenic potential (hypoallergenic)

Dyeability: (two methods) cross-dying method where yarns of one fibre and those of another fibre are woven into a fabric in a desired pattern; solution-dying method provides excellent color fastness under the effects of sunlight, perspiration, air contaminants and washing.

Acetate usually requires dry cleaning.

Production

The Federal Trade Commission definition for acetate fibre is "A manufactured fibre in which the fibre-forming substance is cellulose acetate. Where not less than 92 percent of the hydroxyl groups are acetylated, the term triacetate may be used as a generic description of the fibre."

Acetate is derived from cellulose by deconstructing wood pulp into a purified fluffy white cellulose. In order to get a good product special qualities of pulps - dissolving pulps - are used. A common problem with these is that the reactivity of the cellulose is uneven, and thereby will the quality of the cellulose acetate sometimes be damaged. The cellulose is then reacted with acetic acid and acetic anhydride in the presence of sulfuric acid. It is then put through a controlled, partial hydrolysis to remove the sulfate and a sufficient number of acetate groups to give the product the desired properties. The anhydroglucose unit is the fundamental repeating structure of cellulose and has three hydroxyl groups which can react to form acetate esters. The most common form of cellulose acetate fibre has an acetate group on approximately two of every three hydroxyls. This cellulose diacetate is known as secondary acetate, or simply as "acetate".

After it is formed, cellulose acetate is dissolved in acetone into a viscose resin for extrusion through spinnerets (which resemble a shower head). As the filaments emerge, the solvent is evaporated in warm air via dry spinning, producing fine cellulose acetate fibres.

First U.S. Commercial Acetate Fibre Production: 1924, Celanese Corporation

Current U.S. Acetate Fibre Producers: Celanese Acetate, Eastman Chemical Company

Production Method

- Purified cellulose from wood pulp or cotton linters
- Mixed with glacial acetic acid, acetic anhydride, and a catalyst
- Aged 20 hours- partial hydrolysis occurs
- Precipitated as acid-resin flakes
- Flakes dissolved in acetone
- Solution is filtered

- Spinning solution extruded in column of warm air. Solvent recovered
- Filaments are stretched and wound onto beams, cones, or bobbins ready for use.

Acetate Fibre Characteristics

- cellulosic and thermoplastic
- selective absorption and removal of low levels of certain organic chemicals
- easily bonded with plasticizers, heat, and pressure
- acetate is soluble in many common solvents (especially acetone and other organic solvents) and can be modified to be soluble in alternative solvents, including water
- hydrophilic: acetate wets easily, with good liquid transport and excellent absorption; in textile applications, it provides comfort and absorbency, but also loses strength when wet
- acetate fibres are hypoallergenic
- high surface area
- made from a renewable resource: wood pulp
- can be composted or incinerated
- can be dyed, however special dyes and pigments are required since acetate does not accept dyes ordinarily used for cotton and rayon (this also allows cross-dyeing)
- resistant to mold and mildew
- easily weakened by strong alkaline solutions and strong oxidizing agents.
- can usually be wet cleaned or dry cleaned and generally does not shrink

Major Industrial Acetate Fibre Uses

Apparel: linings, blouses, dresses, wedding and party attire, home furnishings, draperies, upholstery and slip covers.

Industrial uses: cigarette and other filters, ink reservoirs for fibre tip pens.

High absorbency products: diapers and surgical products.

The original Lego bricks were manufactured from cellulose acetate from 1949 to 1963.

Award Ribbon: Rosettes for equestrian events, dog/cat shows, corporate awards, advertising and identification products all use cellulose acetate ribbon.

KEM High End Playing Cards used at the World Series Of Poker & Major Casino Poker rooms are made of Cellulose Acetate.

History of Acetate

Acetate was first introduced in 1904, when Camille Dreyfus and his younger brother Henri, did chemical research and development in a shed in their father's garden in Basle, Switzerland. Inasmuch as their father was interested in a chemical factory, his influence was probably a factor in their choice of careers. And since Basle was a center of the dyestuffs industry, it was natural that their first achievement should be the development of synthetic indigo dyes. In search of a field that offers really limitless potentialities, they deliberately selected that of cellulose acetate products, including fibres for textile use.

For five years, the Dreyfus brothers studied and experimented in a logical, systematic manner in Switzerland and France. By 1910, they had perfected acetate lacquers and plastic film and opened a factory in Basle capable of producing about three tons a day. This was largely sold to the celluloid industry in France and Germany, and to Pathe Fréres in Paris for non-flammable motion picture film base. A small but constantly growing amount of acetate lacquer, called "dope", was sold to the expanding aircraft industry to coat the fabric covering wings and fuselage.

After some twenty-odd thousand separate experiments, by 1913, the brothers produced excellent laboratory samples of acetate continuous filament yarn. The outbreak of the

First World War postponed completion of development leading to successful commercial production until 1921. The war, of course, necessitated rapid expansion of the Basle factory which terminated its trade with Germany and exclusively supplied the Allied Governments with acetate "dope" for military aircraft.

In November 1914, the British Government invited Dr.Camille Dreyfus to come to England to manufacture acetate "dope". In 1917, the War Department of the United States Government invited Dr. Dreyfus to establish a similar factory in the US after their entry into war. After about six weeks, a contract was negotiated for sale of acetate "dope" to the War Department and a plant site was sought. Dr. Dreyfus and his associates started construction of the American company at Cumberland, Maryland in 1918, but the war was over before the plant could be completed. The business with the Government was completed in due time, construction of the plant continued, the early nucleus of the management began to assemble, and the organization in England completed development of the first commercially successful acetate textile yarn. In England, in 1912, the British company produced the first commercial cellulose acetate yarn. The yarn was sold primarily for crocheting, trimming, and effect threads and for popular-priced linings.

The first yarn spun in America was on Christmas Day, 1924, at the Cumberland, Maryland Plant. The first yarn was of fair quality, but sales resistance was heavy, and silk associates worked zealously to discredit acetate and discourage its use. Acetate became an enormous success as a fibre for moire because its thermoplastic quality made the moire design absolutely permanent. The same characteristic also made permanent pleating a commercial fact for the first time, and gave great style impetus to the whole dress industry.

This was a genuine contribution. The mixing of silk and acetate in fabrics was accomplished at the beginning and almost at once cotton was also blended, thus making possible low-cost fabrics by means of a fibre which then was cheaper than silk or acetate. Today, acetate is blended with silk, cotton, wool, nylon, etc. to give to fabrics an excellent wrinkle recovery, good left, handle, draping quality, quick drying, proper dimensional stability, cross-dye pattern potential, at a very competitive price.

Cellulose Triacetate

Cellulose **triacetate**, also known simply as triacetate, is manufactured from cellulose and acetate. Triacetate is typically used for the creation of fibres and film base. It is similar chemically to cellulose acetate, with the distinguishing characteristics being that in triacetate, according to the Federal Trade Commission definition, at least "92 percent of the hydroxyl groups are acetylated." During the manufacture of triacetate the cellulose is completely acetylated whereas in regular cellulose acetate or cellulose diacetate, it is only partially acetylated. Triacetate is significantly more heat resistant than cellulose.

History of Triacetate

Triacetate, whose chemical equation is $[C_6H_7O_2(OOCCH_3)_3]_n$, was first produced commercially in the U.S. in 1954 by Celanese Corporation. Mitsubishi Rayon Co., Ltd. is the only company that still produces the fibre. Eastman Kodak formerly manufactured cellulose triacetate. They discontinued this product on March 15, 2007.

Production

Triacetate is derived from cellulose by combining cellulose with acetate from acetic acid and acetic anhydride. The cellulose acetate is dissolved in a mixture of dichloromethane and methanol for spinning. As the filaments emerge from a spinneret, the solvent is evaporated in warm air, in a process known as dry spinning, leaving a fibre of almost pure triacetate.

A finishing process called S-Finishing or surface saponification is sometimes applied to acetate and triacetate fabrics using a sodium hydroxide solution. This removes part or all of the acetyl groups from the surface of the fibres leaving them with a cellulose coating. This reduces the tendency for the fibres to acquire a static charge.

As a Fibre :

Triacetate fibres have a crenate cross section.

Characteristics :

Shrink resistant

Wrinkle resistant

Easily washable

Generally washable at high temperatures

Maintains creases and pleats well

Usage Scenarios :

Particularly effective in clothing where crease or pleat retention is important, such as skirts and dresses.

In the 1980s triac etate was also used with polyester to create shiny tracksuits which were very popular at the time. The material was very smooth and shiny on the outside and soft and fleecy on the inside.

General Care Tips :

Ironable up to 200 °C

Pleated garments are best hand laundered. Most other garments containing 100% triacetate can be machine washed and dried.

Articles containing triacetate fibres require very little special care due mainly to the fibre's resistance to high temperatures.

As a film

Characteristics :

Resistant to grease, oil, aromatic hydrocarbons, and most common solvents

Films have hard glossy surfaces

Excellent optical clarity

High dielectric constant

Easily laminated, coated, folded, and die-cut

Cellulose acetate film prone to degradation known as Vinegar syndrome

Usage Scenarios:

Polarizer films for LCD projectors

Specialized overhead projector transparencies

Specialized photographic film

Motion picture film

Production of animation cels

Packaging

Copper(II) acetate

Copper(II) acetate, also referred to as cupric acetate, is the chemical compound with the formula $Cu(OAc)_2$ where AcO^- is acetate ($CH_3CO_2^-$). The hydrated derivative, which contains one molecule of water for each Cu atom, is available commercially. $Cu(OAc)_2$ is a dark green crystalline solid, whereas $Cu(OAc)_2(H_2O)_2$ is more bluish-green. Since ancient times, copper acetates of some form have been used as fungicides and green pigments. Today, $Cu(OAc)_2$ is used as a source of copper (II) in inorganic synthesis and as a catalyst or an oxidizing agent in organic synthesis. Copper acetate, like all copper compounds, emits a blue-green glow in a flame.

Structure

The dinuclear structure of copper (II) acetate dihydrate

Copper acetate monohydrate is described by the formula Cu2 (OAc) 4 (H2O) 2. It adopts the "Chinese lantern" structure seen also for related Rh(II) and Cr(II) tetraacetates. One oxygen atom on each acetate is bound to one copper at 1.97 Å (197 pm. Completing the coordination sphere are two water ligands, with Cu-O distances of 2.20 Å (220 pm).

The two five-coordinate copper atoms are separated by only 2.65 Å (265 pm), which is close to the Cu—Cu separation in metallic copper. The two copper centers interact resulting in a diminishing of the magnetic moment such that near 90 K, Cu2 (OAc) 4 (H2O) 2. is essentially diamagnetic due to cancellation of the two opposing spins. Cu2 (OAc) 4 (H2O) 2. was a critical step in the development of modern theories for antiferromagnetic coupling.

History

Copper(II) acetate is occasionally the primary component of verdigris, the blue-green substance that forms on copper during long exposures to atmosphere. It was historically prepared in vineyards, since acetic acid is a byproduct of fermentation. Copper sheets were alternately layered with fermented grape skins and dregs left over from wine production and exposed to air. This would leave a blue substance on the outside of the sheet. This was then scraped off and dissolved in water. The resulting solid was used as a pigment, or combined with arsenic trioxide to form copper acetoarsenite, a powerful insecticide and fungicide called Paris Green or Schweinfurt Green.

Copper (cupric) acetate tablets were believed to repel sharks in the mid twentieth century. Scuba divers strapped tablets of the compound to their belt and/or ankles to provide protection against sharks. It was used by Jacques-Yves Cousteau and his researchers with questionable results.

Synthesis

Copper(II) acetate has been synthesized for centuries by the method described in the history section. This method, however, leads to an impure copper(II) acetate. In a laboratory, a much purer form can be synthesized in a simple three-step procedure. The overall reaction is as follows:

$2\ CuSO_4 \cdot 5H_2O + 4\ NH_3 + 4\ CH_3COOH \rightarrow Cu_2(OAc)_4(H_2O)_2 + 2\ [NH_4]_2[SO_4] + 8\ H_2O$

The hydrate form can be dehydrated by heating at 100 °C in a vacuum:

$Cu_2(OAc)_4(H_2O)_2 \rightarrow Cu_2(OAc)_4 + 2\ H_2O$

Heating a mixture of anhydrous $Cu_2(OAc)_4$ and copper metal affords colorless, volatile cuprous acetate:

$2\ Cu + Cu_2(OAc)_4 \rightarrow 4\ CuOAc$

Uses in chemical synthesis

The uses for copper(II) acetate are more plentiful as a catalyst or oxidizing agent in organic syntheses. For example, $Cu_2(OAc)_4$ is used to couple two terminal alkynes to make a 1,3-diyne:[6]

$Cu_2(OAc)_4$ + 2 RCa"CH → 2 CuOAc + RCa"C-Ca"CR + 2 HOAc

The reaction proceeds via the intermediacy of copper (I) acetylides, which are then oxidized by the copper (II) acetate, releasing the acetylide radical. A related reaction involving copper acetylides is the synthesis of ynamines, terminal alkynes with amine groups using $Cu_2(OAc)_4$.

Sodium Acetate

Sodium acetate, (also sodium ethanoate) is the sodium salt of acetic acid. It is an inexpensive chemical produced in industrial quantities for a wide range of uses.

Applications

Sodium acetate is used in the textile industry to neutralize sulfuric acid waste streams, and as a photoresist while using aniline dyes. It is also a pickling agent in chrome tanning, and it helps to retard vulcanization of chloroprene in synthetic rubber production.

Sodium acetate may be added to foods as a seasoning, and to alcoholic beverages to decrease the risk of a hangover. It may be used in the form of sodium diacetate — a 1:1

complex of sodium acetate and acetic acid, given the E-number E262.

As the conjugate base of a weak acid, a solution of sodium acetate and acetic acid can act as a buffer to keep a relatively constant pH. This is useful especially in biochemical applications where reactions are pH dependent.

Sodium acetate is also used in consumer heating pads or hand warmers and is also used in hot ice. Sodium acetate trihydrate crystals melt at 58 °C, dissolving in their water of crystallization. When they are heated to around 100 °C, and subsequently allowed to cool, the aqueous solution becomes supersaturated. This solution is capable of supercooling to room temperature without forming crystals. By clicking on a metal disc in the heating pad, a nucleation center is formed which causes the solution to crystallize into solid sodium acetate trihydrate again. The bond-forming process of crystallization is exothermic, hence heat is emitted. The latent heat of fusion is about 264–289 kJ/kg. Unlike some other types of heat packs that depend on irreversible chemical reactions, sodium acetate heat packs can be easily recharged by boiling until all crystals are dissolved. Therefore they can be recycled indefinitely.

Preparation

Sodium acetate is very expensive, and is usually purchased from chemical suppliers, instead of being synthesized in the laboratory. It is sometimes produced in a laboratory experiment by the reaction of acetic acid with sodium carbonate, sodium bicarbonate, or sodium hydroxide. These reactions produce aqueous sodium acetate, and water. Carbon dioxide is produced in the reaction with sodium carbonate and bicarbonate, and it leaves the reaction vessel as a gas (unless the reaction vessel is pressurized).

$$CH_3–COOH + Na^+[HCO_3]^- \rightarrow CH_3–COO^- Na^+ + H_2O + CO_2$$

This is the well-known "fizzing" reaction between baking soda and vinegar. 84 grams of sodium bicarbonate (baking soda) react with 750 ml of 8% vinegar to make 82 g sodium acetate in water. By boiling off most of the water, one can refine either a concentrated solution of sodium acetate or crystals.

Reactions

Sodium acetate can be used to form an ester with an alkyl halide such as bromoethane:

$H_3C–COO^- Na^+ + Br–CH_2–CH_3 \rightarrow H_3C–COO–CH_2–CH_3 + NaBr$

In such a reaction, the sodium acetate is usually completed with caesium in order to increase the nucleophilicity of the carboxylate group.

Acetylation

Acetylation (or in IUPAC nomenclature ethanoylation) describes a reaction that introduces an acetyl functional group into an organic compound. Deacetylation is the removal of the acetyl group.

Moreover, it is that process of introducing an acetyl group (resulting in an acetoxy group) into a compound, specifically, the substitution of an acetyl group for an active hydrogen atom. A reaction involving the replacement of the hydrogen atom of a hydroxyl group with an acetyl group (CH_3 CO) yields a specific ester, the acetate. Acetic anhydride is commonly used as an acetylating agent reacting with free hydroxyl groups. For example, it is used in the synthesis of Aspirin.

Acetylation of Proteins

In biology, i.e. in living cells, acetylation occurs as a co-translational and post-translational modification of proteins, for example, histones, p53, and tubulins.

N-alpha-terminal Acetylation

Acetylation of the N-terminal alpha-amine of proteins is a widespread modification in eukaryotes. 40-50% of yeast proteins, and 80-90% of human proteins are modified in this manner, and the pattern of modification is found to be conserved throughout evolution. The modification is performed by N-alpha-acetyltransferases (NATs), a sub-family of the GNAT superfamily of acetyltransferases, which also include histone acetyl transferases. The GNATs transfer the acetyl group from acetyl-coenzyme A to the amine group. The NATs have been most extensively studied in yeast. Here three NAT complexes, NatA/B/C, have been found to perform most N-alpha-terminal acetylations. They have sequence specificity for their substrates, and it is believed that they are associated with the ribosome, where they acetylate nascent polypeptides co-transitionally.

In humans, the human NatA and NatB complexes have been identified and characterized. Subunits of the human NatA complex have been coupled to cancer-related processes such as hypoxia-response and the beta-catenin pathway. It has been found to be over-expressed in papillary thyroid carcinoma and neuroblastoma. The human NatB complex have been coupled to cell cycle. The hnat3 subunit of the hNatB complex have been found over expressed in some forms of cancer.

Despite being such a conserved and widespread modification, little is known about the biological role of N-alpha-terminal acetylation. Proteins such as actin and tropomyosin have been found to be dependent of NatB acetylation to form proper actin filaments. This is yet only an example of the potential importance of this modification.

Lysine Acetylation and Deacetylation

In histone acetylation and deacetylation, the histones are acetylated and deacetylated on lysine residues in the N-terminal tail as part of gene regulation. Typically, these reactions are catalyzed by enzymes with "histone acetyltransferase" (HAT) or "histone deacetylase" (HDAC) activity, although it should be noted that HATs and HDACs can modify the acetylation status of non-histone proteins as well.

The regulation of transcription factors, effector proteins, molecular chaperones, and cytoskeletal proteins by acetylation/deacetylation is emerging as a significant post-translational regulatory mechanism analogous to phosphorylation by the action of kinases or dephosphorylated by the action of phosphatases. Not only can the acetylation state of a protein modify its activity, there has been recent suggestion that this post-translational modification might crosstalk with phosphorylation, methylation, ubiquitination, sumoylation, and others for dynamic control of cellular signaling.

The tubulin acetylation and deacetylation system is well worked out in Chlamydomonas. A tubulin acetyltransferase located in the axoneme acetylates a specific lysine residue in the ±-tubulin subunit in assembled microtubule. Once disassembled, this acetylation can be removed by another specific deacetylase which is cytosolic. Thus the axonemal microtubules (long half life) carry this signature acetylation absent from cytosolic microtubules (short half life).

Histone Acetylation and Deacetylation

In histone acetylation and deacetylation, the histones are acetylated and deacetylated on lysine residues in the N-terminal tail and on the surface of the nucleosome core as part of gene regulation. Typically, these reactions are catalyzed by enzymes with "histone acetyltransferase" (HAT) or "histone deacetylase" (HDAC) activity. The source of the acetyl group in histone acetylation is Acetyl-Coenzyme A, and in histone deacetylation the acetyl group is transferred to Coenzyme A.

Acetylated histones and nucleosomes represent a type of epigenetic tag within chromatin. Acetylation brings in a negative charge, acting to neutralise the positive charge on the histones and decreases the interaction of the N termini of histones with the negatively charged phosphate groups of DNA. As a consequence, the condensed chromatin is transformed into a more relaxed structure which is associated with greater levels of gene transcription. This relaxation can be reversed by HDAC activity. Relaxed, transcriptionally active DNA is referred to as euchromatin. More condensed (tightly packed) DNA is referred to as heterochromatin. Condensation can be brought about by processes including deacetylation and methylation; the action of methylation is indirect and has no effect upon charge.

This charge neutralization model has been challenged by recent studies, according to which transcriptionally active genes are correlated with rapid turnover of histone acetylation. This requires that the HATs and HDACs must act continuously on the affected histone tail. Methylation at a specific lysine residue (K4) is involved in targeting histone tails for continuous acetylation and deacetylation.

Histone acetyltransferases (HAT) are enzymes that acetylate conserved lysine amino acids on histone proteins by transferring an acetyl group from acetyl CoA to form μ-N-acetyl lysine.

Histone acetylation is generally linked to transcriptional activation. These are generally associated with euchromatin. Initially, it was thought that acetylation of lysine neutralizes the positive charge normally present, thus reducing affinity between histone and (negatively charged) DNA which renders DNA more accessible to transcription factors. More recently, it has emerged that lysine acetylation and other posttranslational modifications of histones generate binding sites for specific protein-protein interaction domains, such as the acetyl-lysine binding bromodomain

Interaction with HDACs

Histone acetyltransferases (HATs) and histone deacetylases (HDACs) are recruited to their target promoters through a physical interaction with a sequence-specific transcription factor (TF). They usually function within a multimolecular complex ('enzymatic complex'), in which the other subunits are necessary for them to modify nucleosomes around the binding site. These enzymes can also modify factors other than histones (protein X).

Histone deacetylases (HDAC) (EC number 3.5.1) are a class of enzymes that remove acetyl groups from an μ-N-acetyl lysine amino acid on a histone. Its action is opposite to that of histone acetyltransferase.

Subtypes

HDAC proteins are found in three groups, the first two groups belong to the classical HDACs and their activities are inhibited by trichostatin A (TSA) whereas the third group is a family of NAD+-dependent proteins not affected by TSA. Homologues to all three groups are found in yeast having the names reduced potassium dependency 3 (Rpd3) - corresponds to class 1, histone deacetylase 1 (hda1) – to class 2 and silent information regulator 2(Sir2) – class3.

Subcellular Distribution

Within the class I HDACs, HDAC 1, 2 and 8 are primarily found in the nucleus, whereas HDAC 3 is found both in the nucleus, cytoplasm and also membrane associated. Class II HDACs (HDAC 4, 5, 6, 7 9 and 10) are able to shuttle in and out of the nucelus depending on different signals.

HDAC 6 is a cytoplasmic, microtuble-associated enzyme. HDAC 6 deacetylates tubulin, Hsp90 and cortactin, and forms complexes with other partner proteins and is therefore involved in a variety of biological processes.

Function

Histone tails are normally positively charged due to amine groups present on their lysine and arginine amino acids. These positive charges help the histone tails to interact with and bind to the negatively charged phosphate groups on the DNA backbone. Acetylation, which occurs normally in a cell, neutralizes the positive charges on the histone by changing amines into amides and decreases the ability of the histones to bind to DNA. This process allows chromatin expansion, allowing for genetic transcription to take place. Histone deacetylases remove those acetyl groups, increasing the positive charge of histone tails and encouraging high-affinity binding between the histones and DNA backbone. This process condenses DNA structure, preventing transcription.

Histone deacetylase is involved in a series of pathways within the living system. According to the Kyoto Encyclopedia of Genes and Genomes (KEGG), these are:

Environmental information processing; signal transduction; notch signaling pathway PATH:ko04330.

Cellular processes; cell growth and death; cell cycle PATH:ko04110.

Human diseases; cancers; chronic myeloid leukemia PATH:ko05220.

Histone acetylation plays an important role in the regulation of gene expression. Hyperacetylated chromatin is transcriptionally active, and hypoacetylated chromatin is silent. A study on mice found that a specific subset of mouse genes (7%) was deregulated in the absence of HDAC1. Their study also found a regulatory crosstalk between HDAC1 and HDAC2 and suggest a novel function for HDAC1 as a transcriptional coactivator. HDAC1 expression was found to be increased in the prefrontal cortex of schizophrenia subjects, negatively correlating with the expression of GAD67 mRNA.

It is a mistake to regard HDACs solely in the context of regulating gene transcription by modifying histones and chromatin structure, although that appears to be the predominant function. The function, activity, and stability of proteins can be controlled by post-translational modifications. Protein phosphorylation is perhaps the most widely studied and understood modification in which certain amino acid residues are phosphorylated by the action of protein_kinases or dephosphorylated by the action of phosphatases. The acetylation of lysine residues is emerging as an analogus mechanism, in which non-histone proteins are acted on by acetylases and deacetylases. It is in this context that HDACs are being found to interact with a variety of non-histone proteins — some of these are transcription factors and co-regulators, some are not. Note the following four examples:

HDAC6 is associated with aggresomes. Misfolded protein aggregates are tagged by ubiquitination and removed from the cytoplasm by dynein motors via the microtubule network to an organelle termed the aggresome. HDAC 6 binds polyubiquitinated misfolded proteins and links to dynein motors, thereby allowing the misfolded protein cargo to be physically transported to chaperones and proteasomes for subsequent destruction.

PTEN is an important phosphatase involved in cell signaling via phosphoinositols and the AKT/PI3 kinase pathway. PTEN is subject to complex regulatory control via phosphorylation, ubiquitination, oxidation and acetylation. Acetylation of PTEN by the histone acetyltransferase p300/CBP-associated factor (PCAF) can stimulate its activity; conversely, deacetylation of PTEN by SIRT1 deacetylase and apparently by HDAC1 can repress its activity.

APE1/Ref-1 (APEX) is a multifunctional protein possessing both DNA repair activity (on abasic and single strand break sites) and transcriptional regulatory activity associated with oxidative stress. APE1/Ref-1 is acetylated by PCAF; conversely it is stably associated with and deacetylated by Class I HDACs. The acetylation state of APE1/Ref-1 does not appear to affect its DNA repair activity, but it does regulate its transcriptional activity such as its ability to bind to the PTH promoter and initiate transcription of the parathyroid hormone gene.

NF-kB is a key transcription factor and effector molecule involved in responses to cell stress, consisting of a p50/p65 heterodimer. The p65 subunit is controlled by acetylation via PCAF and by deacetylation via HDAC3 and HDAC6.[13]

These are just some examples of constantly emerging non-histone, non-chromatin roles for HDACs.

HDAC Inhibitors

Histone deacetylase inhibitors (HDIs) have a long history of use in psychiatry and neurology as mood stabilizers and anti-epilectics, for example, valproic acid. More recently, HDIs are being studied as a mitigator or treatment for neurodegenerative diseases. Also in recent years, there has been an effort to develop HDIs for cancer therapy, and Vorinostat (SAHA) has recently been approved for treatment of cutaneous T cell lymphoma (CTCL). The exact mechanisms by which the compounds may work are unclear, but epigenetic pathways are proposed. In addition, a clinical trial is studying valproic acid effects on the latent pools of HIV in infected persons.

HDAC inhibitors have effects on non-histone proteins that are related to acetylation. HDIs can alter the degree of acetylation of these molecules and thereby increase or repress their activity. For the four examples given above (see Function) on HDACs acting on non-histone proteins, in each of those instances the HDAC inhibitor Trichostatin A (TSA) blocks the effect.

Family

Together with the acetylpolyamine amidohydrolases and the acetoin utilization proteins, the histone deacetylases form an ancient protein superfamily known as the histone deacetylase superfamily. Histone deacetylases, acetoin utilization proteins and acetylpolyamine amidohydrolases are members of an ancient protein superfamily

5

Nylon

Man's very first synthetic fibre, stronger than steel, fine as a spider's web, more elastic than any of the common natural fibres, splendidly shiny and due to be marketed with the commercial name of nylon, was presented by Dr. Stine, a DuPont vice president, at the New York World Fair on 27 October 1938. His words, according to the papers, were received with little more than polite respect until he mentioned that a factory had begun to knit nylon hosiery and these new stockings felt better than silk stockings, didn't bag or snag, and when they snagged they didn't run. At this point that passive audience come to see Katherine Hepburn, the famous actress, stood and applauded.

Nylon immediately got off to a flying start. DuPont, indeed, had to build another two plants to flank the first, itself not yet fully worked up. The very first product turned out, even before stockings, was bristles for toothbrushes.

Heavy capital expenditures were needed to industrialise nylon development as well as the production plants.

Nylon stockings quickly become known as nylons. They were presented to the public for the first time at the San Francisco International Exhibition in February 1939 and almost simultaneously made available to the company's Wilmington employees only. In October, the citizens of Wilmington itself were able to buy them. On 15 May 1940 came the great moment when nylons were placed on sale throughout the United States at USD 1.15 - 1.35 a pair. The shops were taken by storm and almost five million pairs were bought on that day alone.

When the United States came into the war a campaign was mounted to collect nylon and nylons for the war effort. Nylon was quickly employed for parachutes fabric, brained parachute cord, outwear and tenting and tire yarn and glider towrope, etc. The vanguard of the U.S. Army floated to earth in Normandy carried by and covered with nylon. During the war some 13,000 tonnes of nylon were produced per year. In 1945, nylons came back to the shops, though about a couple of years were needed to catch up with the stocking demand. By 1949, production had raised to 25,000 tonnes a year.

About Nylon

The origin of the word nylon has been told in what may be called an official story. But as in all stories, there is a legend. One apocryphal versions were passed on more by word of mouth than anything else and are still better known and more widely believed than the official story.

Throughout its development and prior to its introduction on the market, polyamide 66 was simply referred to as fibre 66. When the time came to launch it, a trade name obviously had to be found. One year before nylon announcement, Dr. H. Church, head of research at DuPont's rayon Department at Buffalo, suggested playfully the acroym Duparooh, which stood for DuPont "DuPont ulls a Rabbit Out of hat". Names such as novasilk and synthesilk were discarded because the company wanted its new synthetic fibre to conquer the market on account of all its high qualities and not just as a substitute for silk. A committee of three formed in 1938 collected a list of 400 names but no one of them met approval. Dr. E.K.Gladding, one of the committee member, proposed Norun with stockings in mind, but changed it to Nuron since also stockings of the new fibre would run. Here, however, the question of how this should be pronounced arose. An American "noo" (= new) as in "noon" was possible, but a British "new" would have seemed like "neuron". Replacement of the "r" by an "l" to form "nulon" was proposed, but once again the sound of the "u" would be uncertain and an expression such as "new nulon" would also be cacophonic. At this point a vowel change was suggested: "nilon" instead of "nulon". But was this "ni" to be pronounced as in "need" or in "nine" ? In the end, the name selection committee opted for a "y". The pronunciation problem was solved and "nylon" was born.

This is the official DuPont version. The legend is much more intriguing and presumably offers one of the many examples of a tendency among English-speaking peoples to invent humorous, sarcastic or derogatory versions of familiar acronyms. It was not very difficult, therefore, for some wag to come up with "Now You Lose Old Nippon" or "Now You Lousy Old Nippon". Apart from the evident threat posed by nylon to Japan as the largest producer of natural silk and hence the end of its leadership in the manufacture of fibres and choice fabrics, the period when nylon appeared was marked by very strong anti-Japanese feelings. An American success was thus accompanied by an open desire to offend Japan. The two appealing versions probably saw the light at the same time as nylon itself in October 1938 and quickly spread from coast to coast. Such was their effect, indeed, and so much better were they known than the official version, that in February 1941 DuPont commissioned a Japanese newspaper to publish a denial that the word nylon had the meanings attributed to it. This however has never become sufficiently known to dethrone the legendary versions in the eyes of the mass of mankind, even today.

Wallace Hume Carothers (b. April 27, 1896, d. April 29, 1937) can be considered the father of the science of man-made polymers and the man responsible for the invention of nylon and neoprene. The man was a brilliant chemist, inventor and scholar and a troubled soul. Despite an amazing career, Wallace Carothers held more than fifty patents; the inventor ended his own life.

Wallace Carothers was born in Iowa and first studied accounting and later studied science (while teaching accounting) at Tarkio College in Missouri. While still an undergraduate student, Wallace Carothers became the head of the chemistry department. Wallace Carothers was talented in chemistry but the real reason for the appointment was a personnel shortage due to the war effort (WWI). He received both a Master's degree and PhD from the University of Illinois and then became a professor at Harvard, where he started his research into chemical structures of polymers in 1924.

In 1928, the DuPont chemical company opened a research laboratory for the development of artificial materials, deciding that basic research was the way to go — not a common path for a company to follow at the time. Wallace Carothers left Harvard to lead Dupont's research division. A basic lack of knowledge of polymer molecules existed when Wallace Carothers began his work there. Wallace Carothers and his team were the first to investigate the acetylene family of chemicals.

In 1931, DuPont started to manufacture neoprene, a synthetic rubber created by Carothers' lab. The research team then turned their efforts towards a synthetic fibre that could replace silk. Japan was the United States' main source of silk, and trade relations between the two countries were breaking apart. By 1934, Wallace Carothers had made significant steps toward creating a synthetic silk by combining the chemicals amine, hexamethylene diamine and adipic acid to create a new fibre formed by the polymerizing process and known as a condensation reaction. In a condensation reaction, individual molecules join with water as a byproduct. Wallace Carothers refined the process (since the water produced by the reaction was dripping back into the mixture and weakening the fibres) by adjusting the equipment so that the water was distilled and removed from the process making for stronger fibres.

Extract from "Fortune Magazine" about nylon circa 1938: "nylon breaks the basic elements like nitrogen and carbon out of coal, air and water to create a completely new molecular structure of its own. It flouts Solomon. It is an entirely new arrangement of matter under the sun, and the first completely new synthetic fibre made by man. In over four thousand years, textiles have seen only three basic developments aside from mechanical mass production: mercerized cotton, synthetic dyes and rayon. Nylon is a fourth."

In 1936, Wallace Carothers married Helen Sweetman, a fellow employee at DuPont. They had a daughter, but tragically Wallace Carothers committed suicide before the birth of this first child. It was likely that Wallace Carothers was a severe manic-depressive, and the untimely death of his sister in 1937 added to his depression. A fellow Dupont researcher, Julian Hill, had once observed Carothers carrying what turned out to be a ration of the poison cyanide. Hill remarked that Carothers could list all the famous chemists who had committed suicide. In April of 1937, Wallace Hume Carothers consumed that ration of poison himself and added his own name to that list.

Nylon, the miracle fibre, was introduced to the world in 1938.

Chemistry

Nylons are condensation copolymers formed by reacting equal parts of a diamine and a dicarboxylic acid, so that peptide bonds form at both ends of each monomer in a process analogous to polypeptide biopolymers. Chemical elements included are carbon, hydrogen, nitrogen, and oxygen. The numerical suffix specifies the numbers of carbons donated by the monomers; the diamine first and the diacid second. The most common variant is nylon 6-6 which refers to the fact that the diamine (hexamethylene diamine) and the diacid (adipic acid) each donate 6 carbons to the polymer chain. As with other regular copolymers like polyesters and polyurethanes, the "repeating unit" consists of one of each monomer, so that they alternate in the chain. Since each monomer in this copolymer has the same reactive group on both ends, the direction of the amide bond reverses between each monomer, unlike natural polyamide proteins which have overall directionality: C terminal '! N terminal. In the laboratory, nylon 6-6 can also be made using adipoyl chloride instead of adipic. It is difficult to get the proportions exactly correct, and deviations can lead to chain termination at molecular weights less than a desirable 10,000 daltons (u). To overcome this problem, a crystalline, solid "nylon salt" can be formed at room temperature, using an exact 1:1 ratio of the acid and the base to neutralize each other. Heated to 285 °C, the salt reacts to form nylon polymer. Above 20,000 daltons, it is impossible to spin the chains into yarn, so to combat this, some acetic acid is added to react with a free amine end group during polymer elongation to limit the molecular weight. In practice, and especially for 6,6, the monomers are often combined in a water solution. The water used to make the solution is evaporated under controlled conditions, and the increasing concentration of "salt" is polymerized to the final molecular weight.

DuPont patented nylon 6,6, so in order to compete, other companies (particularly the German BASF) developed the homopolymer nylon 6, or polycaprolactam — not a condensation polymer, but formed by a ring-opening polymerization (alternatively made by polymerizing aminocaproic acid). The peptide bond within the caprolactam is broken with the exposed active groups on each side being incorporated into two new bonds as the monomer becomes part of the polymer backbone. In this case, all amide bonds lie in the same direction, but the properties of nylon 6 are sometimes indistinguishable from those of nylon 6,6 — except for melt temperature (N6 is lower) and some fibre properties in products like carpets and textiles. There is also nylon 9.

Nylon 5,10, made from pentamethylene diamine and sebacic acid, was studied by Carothers even before nylon 6,6 and has superior properties, but is more expensive to make. In keeping with this naming convention, "nylon 6,12" (N-6,12) or "PA-6,12" is a copolymer of a 6C diamine and a 12C diacid. Similarly for N-5,10 N-6,11; N-10,12, etc. Other nylons include copolymerized dicarboxylic acid/diamine products that are not based upon the monomers listed above. For example, some aromatic nylons are polymerized with the addition of diacids like terephthalic acid ('! Kevlar) or isophthalic acid ('! Nomex), more commonly associated with polyesters. There are copolymers of N-6,6/N6; copolymers of N-6,6/N-6/N-12; and others. Because of the way polyamides are formed, nylon would seem to be limited to unbranched, straight chains. But "star" branched nylon can be produced by the condensation of dicarboxylic acids with polyamines having three or more amino groups.

A molecule of water is given off and the nylon is formed. Its properties are determined by the R and R' groups in the monomers. In nylon 6,6, R' = 6C and R = 4C alkanes, but one also has to include the two carboxyl carbons in the diacid to get the number it donates to the chain. In Kevlar, both R and R' are benzene rings.

Nylon Fibre

The **Federal Trade Commission**'s definition for Nylon Fibre: A manufactured fibre in which the fibre forming substance is a long-chain synthetic polyamide in which less than 85% of the amide-linkages are attached directly (-CO-NH-) to two aliphatic groups.

- A synthetic thermoplastic fibre (Nylon melts/glazes easily at relatively low temperatures).
- Round, smooth, and shiny **filament fibres**.
- cross sections can be either.
- trilobal to imitate silk.
- multilobal to increase staple like appearance and hand
- Its most widely used structures are **multifilament**, **monofilament**, **staple** or **tow** and is available as partially drawn or as finished filaments.
- Regular nylon has a round cross section and is perfectly uniform. The filaments are generally completely transparent unless they have been delustered or solution dyed. Thus, they are microscopically recognized as glass rods.
- Molecular chains of nylon are long and straight variations but have no side chains or linkages.
- Cold drawing (step 18 on the model) can align the chains so they are oriented with the lengthwise direction and are highly crystalline.
- Nylon is related chemically to the protein fibres silk and wool.
- They both have similar dye sites but nylon has many fewer dye sites than wool.

Fibre types are produced commercially in various parts of the world. Nylon 66 has been preferred in North American markets, whereas nylon 6 is much more popular in Europe and elsewhere. Nylon is produced by melt spinning and is available in staple, tow, monofilament, and multi-filament form. The fibre has outstanding durability and excellent physical properties. Nylons are semi-crystalline polymers. The amide group -(-CO-NH-)- provides hydrogen bonding between polyamide chains, giving nylon high strength at elevated temperatures, toughness at low temperatures, combined with its other properties, such as stiffness, wear and abrasion resistance, low friction coefficient and good chemical resistance. These properties have made nylons the strongest of all man-made fibres in common use. Because nylons offer good mechanical and thermal properties, they are also a very important engineering thermoplastic. For example, 35% of total nylon produced is used in the automobile industry. There are several commercial nylon products, such as nylon 6, 11, 12, 6/6, 6/10, 6/12, and so on. Of these, the most widely used nylon products in the textile industry are formed of nylon 6 and nylon 6/6. The others are mainly used in tubing extrusion, injection molding, and coatings of metal objects.

Nylon's outstanding characteristic in the textile industry is its versatility. It can be made strong enough to stand up under the punishment tire cords must endure, fine enough for sheer, high fashion hosiery, and light enough for parachute cloth and backpacker's tents. Nylon is used both alone and in blends with other fibres, where its chief contributions are strength and abrasion resistance. Nylon washes easily, dries quickly, needs little pressing, and holds its shape well since it neither shrinks nor stretches.

Nylon Fibre Formation

One of the most important factors in polymer processing is viscosity, which is a function of molecular weight. The number-average molecular weight of polymer suitable for textile fibre production ranges from 14,000 to 20,000. Since Polycaprolactam can be regarded at equilibrium as a polycondensation polymer, the number-average molecular weight alone is sufficient for its characterization. Two-step melt spinning, comprised of spinning and drawing, is considered to be the conventional method to manufacture nylon filaments. After melting, filtering, and deaerating, the molten polymer is extruded through a spinneret into a chamber where the melt solidifies into a filament form. At this stage, the filaments have little molecular orientation, and their slight birefringence is due to shear forces set up during extrusion. In order to achieve desirable properties through molecular orientation and crystallinity, the newly formed filaments must be drawn. Since the Tg of nylon is below room temperature, nylon can be cold drawn.Hot drawing is also frequently used. Nylon filaments are drawn approximately four times their initial length. The effect of drawing on birefringence, a measure of molecular anisotropy, can be seen in Table I. Also, the elastic modulus increases significantly with increasing orientation as shown in Table I. Other physical properties, such as density equilibrium, moisture sorption, tenacity and elongation-at-break, are also affected by drawing.

Table 1: Effect of Drawing on Birefringence and Elastic modulus

Draw Ratio	Birefringence	Elastic Modulus (GPa)
1.	0.00832	1.97
2.	0.03297	2.74
3.	0.05523	3.70
4.	0.05904	4.59
5.	0.06381	5.77
6.	0.06901	6.74

Rather than two-step spinning (extrusion) and drawing, a one-step, high-speed spinning process is being used increasingly. In high-speed spinning, filament windup speed relative to the extrusion speed is very high and orientation and crystallization occur in elongation flow along the spin line. When drawing as-spun fibres, the molecules are arranged randomly in amorphous regions and as folded chains in crystalline region

In essence, cold drawing stretches chains in amorphous regions, but molecular folds are restricted and the molecules orient themselves along the fibre axis direction, resulting in enhanced orientation and high crystallinity. In the case of nylons, which have sheet-like crystal structures, drawing may enable the hydrogen-bonded polyamide sheets to slip past each other and form more oriented structure. Hot drawing is a procedure using high temperature during drawing and annealing under restraint after drawing. Exposure to high temperature helps to increase the draw ratio, and higher moduli and tenacity can be achieved.. Ultra drawing of solidified crystalline material induces a high degree of chain extension which leads to very high tensile strength and modulus. This results in a so-called high-performance fibre.

A skin-core structure, mostly depending on spinning speed, is generally formed within melt-spun fibres. At a constant feeding rate, higher spinning speeds will produce more extended chains in the melt and form a finer filament. Therefore, the finer fibre usually has higher modulus and tenacity. Fine filament cannot be drawn as much as a coarse filament, because partial orientation on the outer parts of the filaments is formed when the molten fluid is drawn over the sides of the orifice. As a result, finer filaments have a greater proportion of 'skin' to bulk, i.e., better orientation has already been formed. Naturally, there is not much space for an improvement by cold drawing within fine filaments. The filaments become lustrous and strong.

Features

Nylon shares the properties of organic polymer molecules. Chemically, nylon is formed by chains of amide molecules. The chains are arranged parallel to each other, attached by hydrogen bonds. To make nylon's fibres strong, a polymerizing process must occur that allows the molecules to combine without retaining any water. In effect the process of making nylon is a condensation reaction.

Function

Chemically, nylon fibres are linear polyamides which are extremely versatile. Nylon is a very strong but lightweight material. It stretches but also regains its original shape easily. Nylon can be dyed almost any color and usually is a bit shiny. It is difficult to tear or damage, lasts a long time, and often can be machine washed and dried.

Significance

Wallace Carothers invented nylon. He had been hired by the E.I. dupont de Nemous and Company when it decided to open a research laboratory in 1928. This was an innovative move on the part of the company but it was motivated, in part, by the cost of importing silk from Japan in this period between the world wars. Before going to dupont, Carothers had taught organic chemistry at Harvard

Rheological Behaviour

The melt viscosity of the polymer can be represented as a function of molecular weight by the relationship:

$$\cdot = K\ (M_w)^a$$

Where is the zero shear viscosity, M_w it the weight average molecular weight, K and an are constants dependent upon the polymer and temperature. In the case of nylons, the value of exponent a normally is in the range of 3.4-3.8.

It has long been known that moisture has a strong effect on the rheological behavior of nylons. Generally, high moisture levels cause degradation and foaming, and relatively low levels of moisture act as plasticizer in nylon 6 during melt processing. All nylons absorb moisture. The extent of moisture absorption depends on temperature, crystallinity, and humidity. Therefore, before processing of nylon resins, the polymer pellets must be dried to moisture levels below 0.2 wt%, in order to avoid bubble formation and significant polymer degradation during processing. A recent study found that the drying temperatures used affect zero shear melt viscosity. The result is shown in

Table 2. Effect of Resin Drying Temperature on Zero Melt Viscosity

Drying Temperature (°C)*	Molecular Weight Exponent
50	3.8
110	4.6-5.4

Non-conventional Spinning Techniques

Alternative to conventional melt spinning, various solution-spinning techniques have been introduced. Solution spinning techniques (gel, wet, dry) enable the spinning of high molecular weight polyamides, leading to high tenacity filaments (tenacity 100cN/tex). As an innovation on fibre formation, new technologies producing micro fibres have been developed and reported. Primarily direct spinning and mechanical and solvent splitting produce micro fibres. Electro spinning represents another approach to fibre spinning, when electrical forces on polymer melt or solution surface overcome the surface tension and cause an ejection from an electrically charged jet. The diameter of the fibres produced by this technique is of the order of nanometers. Frequently, there are produced fibres that are electrically charged.

Crystalline structure : Both nylon 6 and nylon 66 are semi-crystalline polymers. These linear aliphatic polyamides are able to crystallize mostly because of strong intermolecular hydrogen bonds through the amide groups and because of Vander walls forces between the methylene chains. Since these unique structural and thermo-mechanical properties of nylons are dominated by the hydrogen bonds in these polyamides, quantum chemistry can be used to determine the hydrogen bond potential. It shows hydrogen-bonding planes, and the right side shows the view down the chain axis. For the -form of nylon 6, adjacent chains are ant parallel and the hydrogen bonding is between adjacent chains within the same sheet (bisecting the CH_2 angles). For the -form of nylon 6, the chains are parallel and the hydrogen bonding is between chains in adjacent sheets. . In nylon 66, the chains have no directionality. Research results have shown that the stable crystalline structure is the -form comprised of stacks of planar sheets of hydrogen-bonded extended chains. It also appears that Young's modulus of the α -form is higher than the ³-form.

Mechanical, thermal and optical properties of fibres are strongly affected by orientation and crystallinity. Basically, higher fibre orientation and crystallinity will produce better properties. Crystallinity of nylons can be controlled by nucleation, i.e., seeding the molten polymer to produce uniform sized smaller spherulites. This results in increased tensile yield strength, flexural modulus, creep resistance, and hardness, but some loss in elongation and impact resistance. Another important benefit obtained from nucleation is decrease of setup time during processing.

Dyeability

The dyeing efficiency of nylon fibres is enhanced due to the end groups -COOH and -NH2, which exhibit polar and hydrophilic characteristics. Dye diffusion into fibres is closely related to the rate of dyeing, level of dyeing through dye migration, wet fastness properties of dyes, etc. It is generally believed that dye diffusively is independent on dye concentration, with some exception. T. Shibusawa studied the diffusion of most disperse dyes on nylon 6 and found that the actual diffusively on nylon 6 fibres is not always independent on dye concentration. Kim et al. have reported that both dyeing rate and dye saturation of 1,4 -diaminoanthraquinone (1,4-DAA) were improved considerably in the presence of didodecyldimethlammonium bromide (DDDMAB). The amount of DDDMAB adsorbed on nylon 6 fibre is roughly 20 times higher than that a conventional dispersing agent. This suggests that there might be fairly strong interaction between DDDMAB and the fibre by virtue of electrostatic and hydrophobic interactions. There have been many attempts to improve nylon's dyeability or at least to point out the factors and mechanisms acting in nylon dyeing. It has been shown that acrylonitrile and styrene radiation grafting on the polymer could improve the dyeability of nylon. Another approach to higher dyeability of nylon 6 is by copolymerization. In this case, the dyeability can be improved at the expense of a decrease of specific viscosity and of heat and hydrolysis resistance. Other treatments, such as plasma etching and superheated steaming have proved to decrease nylon dyeability. In the former treatment, outer structures, not normally susceptible to dyes, are etched away whereas the crystalline phases inside the fibre are not as much affected. Superheated steaming of the fibres leads to higher shrinkage and to higher crystallinity and crystal size, which contribute to decrease dyeability.

Disperse Dyes

The general structure of a polyester is shown below:

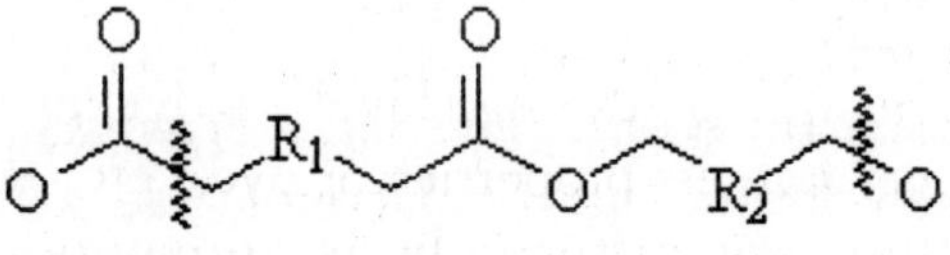

They tend to be fairly hydrophobic (though this depends on the structure of R_1 and R_2), but not as much as, say, a long hydrocarbon would be, since the -COOC- groups cause some polarity. The polymer chains in a sample of polyester are highly crystalline (for a polymer) and quite tightly packed together. The result of this is that polyesters have very little affinity for large ionic dyes- the dyes simply cannot either distribute between the chains, or form satisfactory intermolecular interactions. Therefore, acid and direct dye classes are useless for this polymer.

Disperse dyes have low solubility in water, but they can interact with the polyester chains by forming dispersed particles. Their main use is the dyeing of polyesters, and they find minor use dyeing cellulose acetates and polyamides. The general structure of disperse dyes is small, planar and non-ionic, with attached polar functional groups like $-NO_2$ and -CN. The shape makes it easier for the dye to slide between the tightly-packed polymer chains, and the polar groups improve the water solubility, improve the dipolar bonding between dye and polymer and affect the colour of the dye. However, their small size means that disperse dyes are quite volatile, and tend to sublime out of the polymer at sufficiently high temperatures.

The dye is generally applied under pressure, at temperatures of about 130°C. At this temperature, thermal agitation causes the polymer's structure to become looser and less crystalline, opening gaps for the dye molecules to enter. The interactions between dye and polymer are thought to be Van-der-Waals and dipole forces.

The volatility of the dye can cause loss of colour density, and staining of other materials at high temperatures. This can be counteracted by using larger molecules, or making the dye more polar (or both). This has a drawback, however, in that this new larger, more polar molecule will need more extreme forcing conditions to dye the polymer.

Classes of Disperse Dye

The most important class is the azo class. This class of azo disperse dyes may be further sub-divided into four groups, the most numerous of which is the aminoazobenzene class. This class of dye can be altered as mentioned before, to produce bathochromic shifts. A range of heterocyclic aminoazobenzene dyes are also available. These give bright dyes, and are bathochromically shifted to give blues. The third class of disperse dye is based on heterocyclic coupling components, which produce bright yellow dyes. The fourth class are disazo dyes. These tend to be quite simple in structure. Other than these, there are disperse dyes of the carbonyl class, and a few from the nitro and polymethine classes. Below is an example of a disperse dye. It is the same as the chime molecule at the top of the page.

Basic Dyes

Basic dyes possess cationic functional groups such as -NR_3^+ or =NR_2^+. The name 'basic dye' refers to when these dyes were still used to dye wool in an alkaline bath. Protein in basic conditions develops a negative charge as the -COOH groups are deprotonated to give -COO^-. Basic dyes perform poorly on natural fibres, but work very well on acrylics.

A general structure of an acrylic type polymer is shown below. It is simplified, and doesn't show any anionic groups which are often present.

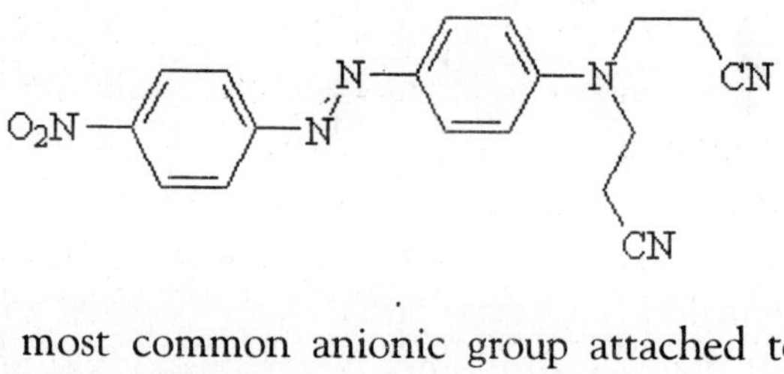

The most common anionic group attached to acrylic polymers is the sulphonate group, -SO_3^-, closely followed by the carboxylate group, -CO_2^-. These are either introduced as a result of co-polymerisation, or as the residues of anionic polymerisation inhibitors. It is this anionic property which makes acrylics suitable for dyeing with cationic dyes, since there will be a strong ionic interaction between dye and polymer (in effect, the opposite of the acid dye-protein fibre interaction). An example of a basic dye is shown below:

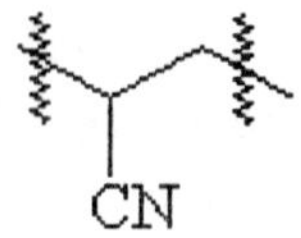

In this case, the charge is delocalised across the molecule, but in some basic dyes, the charge is located on a single nitrogen atom.

The structure of nylon 6,6 and nylon 6 are shown below:

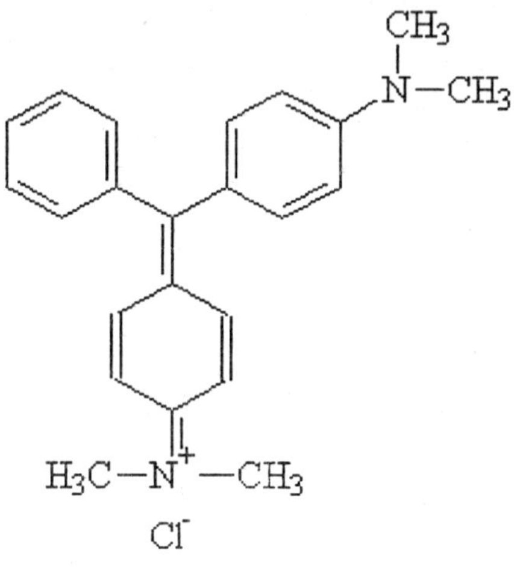

It can be seen that there is a distinct similarity between these polymers and that of keratin (the nylons lack R groups, and have more space between each amide group). The dyeing of these polymers is quite similar to the acid dye action on wool.

Degradation

The -COOH and $-NH_2$ end-groups in nylons are sensitive

to light, heat, oxygen, acids and alkali. When exposed to elevated temperatures, unmodified nylons undergo molecular weigh degradation, which results in loss of mechanical properties. The degradation is highly time/temperature dependent. By adding heat stabilizer, nylon can be used at elevated temperature for long-term performance. Exposure to UV light results in degradation nylon over an extended period of time, it appears that adding carbon black can reduce the radiation degradation. Nylons are chemical resistance to hydrocarbons, aromatic and strong acids, bases, and phenols attack aliphatic solvents, but them. They also are gradually attacked hydrolytically by hot water. Newly developed sulfonation of nylon 6 fibre by 2,5 dichlorobenzene sulfonyl chloride (DSBC) has a great effect on the heat and chemical stability of the fibres. It reported that the modified fibre is non-melting up to 1000°C, and does not burn when put it in direct flame (but chars without losing fibre form). It does not dissolve in formic acid and concentrated mineral acid. Its glass transition temperature is about 500°C.

Non wovens Usage

The fibre has outstanding durability and excellent physical properties. Like PET fibre, it has a high melting point, which conveys good high- temperature performance. The fibre is more water sensitive than PET; despite this fact, nylon is not considered a comfortable fibre in contact with the skin. Its toughness makes it a major fibre of choice in carpets, including needle punched floor-covering products. Because of its relatively high cost, nylon has somewhat limited use in nonwoven products. It is used as a blending fibre in some cases, because it conveys excellent tear strength. The resiliency and wrinkle recovery performance of a nonwoven produced from nylon is not as excellent as that from PET fibre.

World Consumption of Nylon Fibre in Nonwovens 1998- 2007

It is forecasted that global consumption of nonwoven may reach 3.7 million tons by 2005 and 4 million tones by 2007. Consumption of manmade fibre was about 8.1% of all Textile fibres in 1998. In 2005 it is expected to reach 10% and 10.4 % by 2007. Table 3 shows the forecasted consumption of nylon fibre in nonwoven Industry till 2007.

Year	Consumption in thousand tons
1998	36
1999	39
2000	49
2001	50
2005	56
2007	60

In certain applications, the performance of nylon fibre is hard to beat. However, because of its higher cost, it is used in specialized applications where its performance can justify the increased cost. It is used as a blending fibre in some cases, because it conveys excellent tear strength. The resiliency and wrinkle recovery performance of a nonwoven produced from nylon is not as excellent as that from PET fibre. This polymer is used in moderate quantities, because it is more expensive than polyester, polypropylene, or rayon. Some particular applications are as follows:

It can be mostly found in garment interlinings and wipes where it supplies strength and resilience.

In Ni/H and Ni/Cd batteries, nylon fibres are used as Nonwovens separators.

Nylon fibres are used for the manufacture of split table-pie fibres. These fibres find application in high performance wipes, synthetic suede, heat insulators, battery separators and specialty papers.

Nonwovens developed from nylon are found in automotive products, athletic wear and conveyor belts.

Basic Concepts of Nylon Production

The first approach: combining molecules with an acid (COOH) group on each end are reacted with two chemicals that contain amine (NH_2) groups on each end.

This process creates **nylon 6,6**, made of hexamethylene diamine with six carbon atoms and acidipic acid, as well as six carbon atoms.

The second approach: a compound has an acid at one end and an amine at the other and is polymerized to form a chain with repeating units of ($-NH-[CH_2]n-CO-$)x.

In other words, nylon 6 is made from a single six-carbon substance called caprolactam.

In this equation, if n=5, then **nylon 6** is the assigned name. (may also be referred to as polymer)

Nylon 6,6

- Pleats and creases can be heat-set at higher temperatures
- Nylon 6 is very easy to dye, but Nylon 6,6 is not

Nylon 6

- Better dye Affinity
- Softer Hand
- Greater elasticity and elastic recovery
- Better weathering properties; better sunlight resistance

Producers

The producers of nylon include: Honeywell Nylon Inc., Invista, Wellman Inc. among many others. The DuPont Company, is the most famous pioneer of the nylon we know today.

Characteristics

- Variation of luster: nylon has the ability to be very lustrous, semilustrous or dull.

- Durability: its high tenacity fibres are used for seatbelts, tire cords, ballistic cloth and other uses.
- High elongation.
- Excellent abrasion resistance.
- Highly resilient (nylon fabrics are heat-set).
- Paved the way for easy-care garments.
- High resistance to:
- Insects, fungi and animals.
- Molds, mildew, rot.
- many chemicals.
- Used in carpets and nylon stockings.
- Melts instead of burning.
- Used in many military applications.

Bulk Properties

Above their **melting temperatures**, T_m, **thermoplastics** like nylon are **amorphous solids** or viscous **fluids** in which the chains approximate **random coils**. Below T_m, amorphous regions alternate with regions which are **lamellar crystals**. The amorphous regions contribute elasticity and the crystalline regions contribute strength and rigidity. The **planar** amide (-CO-NH-) groups are very **polar**, so nylon forms multiple **hydrogen bonds** among adjacent strands. Because the nylon backbone is so regular and symmetrical, especially if all the amide bonds are in the **trans configuration**, nylons often have high crystallinity and make excellent fibres. The amount of crystallinity depends on the details of formation, as well as on the kind of nylon. Apparently it can never be **quenched** from a **melt** as a completely amorphous solid.

Nylon 6,6 can have multiple parallel strands aligned with their neighboring peptide bonds at coordinated separations of exactly 6 and 4 carbons for considerable lengths, so the **carbonyl oxygens** and amide **hydrogens** can line up to form interchain **hydrogen bonds** repeatedly, without interruption.

Nylon 5,10 can have coordinated runs of 5 and 8 carbons. Thus parallel (but not antiparallel) strands can participate in extended, unbroken, multi-chain ²-**pleated sheets**, a strong and tough supermolecular structure similar to that found in natural **silk fibroin** and the ²-**keratins** in **feathers**. (Proteins have only an amino acid ±-carbon separating sequential -CO-NH- groups.) Nylon 6 will form uninterrupted **H-bonded** sheets with mixed directionalities, but the ²-sheet wrinkling is somewhat different. The three-dimensional disposition of each **alkane hydrocarbon chain** depends on **rotations** about the 109.47° **tetrahedral** bonds of singly-bonded carbon atoms.

When **extruded** into fibres through pores in an **industrial spinneret**, the individual polymer chains tend to align because of **viscous flow**. If subjected to **cold drawing** afterwards, the fibres align further, increasing their crystallinity, and the material acquires additional **tensile strength**. In practice, nylon fibres are most often drawn using heated rolls at high speeds.

Block nylon tends to be less crystalline, except near the surfaces due to **shearing stresses** during formation. Nylon is **clear** and **colorless**, or milky, but is easily **dyed**. Multistranded nylon cord and rope is slippery and tends to unravel. The ends can be **melted** and fused with a heat source such as a **flame** or **electrode** to prevent this.

When dry, polyamide is a good electrical insulator. However, polyamide is **hygroscopic**. The absorption of water will change some of the material's properties such as its **electrical resistance**. Nylon is less absorbent than wool or cotton.

Historical Uses

Bill Pittendreigh, **DuPont**, and other individuals and corporations worked diligently during the first few months of **World War II** to find a way to replace Asian **silk** and **hemp** with nylon in **parachutes**. It was also used to make **tires**, **tents**, **ropes**, **ponchos**, and other **military** supplies. It was even used in the production of a high-grade paper for U.S. **currency**.

At the outset of the war, **cotton** accounted for more than 80% of all fibres used and manufactured, and **wool** fibres accounted for the remaining 20%. By August 1945, manufactured fibres had taken a market share of 25% and cotton had dropped.

Some of the terpolymers based upon nylon are used every day in packaging. Nylon has been used for **meat** wrappings and **sausage** sheaths.

Use in Composites

Nylon can be used as the matrix material in **composite materials**, with reinforcing fibres like glass or carbon fibre, and has a higher **density** than pure nylon. Such thermoplastic composites (25% glass fibre) are frequently used in car components next to the engine, such as intake manifolds, where the good heat resistance of such materials makes them feasible competitors to metals.

Hydrolysis and Degradation

All nylons are susceptible to hydrolysis, especially by **strong acids**, a reaction essentially the reverse of the synthetic reaction shown above. The **molecular weight** of nylon products so attacked drops fast, and cracks form quickly at the affected zones. Lower members of the nylons (such as nylon 6) are affected more than higher members such as nylon 12. This means that nylon parts cannot be used in contact with **sulphuric acid** for example, such as the electrolyte used in **lead-acid batteries**. When being molded, nylon must be dried to prevent **hydrolysis** in the molding machine barrel since water at high temperatures can also degrade the polymer.

Nylon Quick Facts:

- Maximum Temperature: 210°F 99°C
- Minimum Temperature: -94°F -70°C
- Autoclavable: No
- Melting Point: 420°F 216°C

- Tensile Strength: 5,800 psi
- Hardness: R92
- UV Resistance: Good
- Translucent
- Rigid
- Specific Gravity: 1.13

Nylon Resistance:

- Excellent resistance (no attack) to Oils, Bases and THF
- Good resistance (no attack) to Solvents, Formaldehyde and Alcohols
- Limited resistance (moderate attack and suitable for short term use only) to Dilute Acids
- Poor resistance (not recommended for use with) Phenols, Alkalis, Iodine and Acids

Nylon Fabrication:

- Excellent material for machining
- Tough, strong, and impact resistant material
- Very low coefficient of friction
- Abrasion resistant

Nylon 6

Nylon 6 or polycaprolactam is a **polymer** developed by **Paul Schlack** at **IG Farben** to reproduce the properties of **nylon 6,6** without violating the **patent** on its production. Unlike most other **nylons**, nylon 6 is not a **condensation polymer**, but instead is formed by **ring-opening polymerization.** This makes it a special case in the comparison between condensation and **addition polymers.** Its **competition** with nylon 6,6 and the example it set have also shaped the economics of the **synthetic fibre** industry. It was given the **trademark** Perlon in the year 1952.

Synthesis

Nylon 6 begins as pure **caprolactam**. As caprolactam has 6 carbon atoms, it got the name Nylon-6.

When caprolactam is heated at about 533 K in an inert **atmosphere of nitrogen** for about 4-5 **hours**, the ring breaks and undergoes **polymerization**. Then the molten mass is passed through spinnerets to form fibres of Nylon 6.

During **polymerization**, the **peptide** bond within each caprolactam **molecule** is broken, with the active groups on each side re-forming two new bonds as the **monomer** becomes part of the polymer backbone. Unlike nylon 6,6, in which the direction of the **amide** bond reverses at each bond, all nylon 6 amide bonds lie in the same direction. Nylon 6 therefore resembles natural poly**peptides** more closely; in fact, caprolactam would become an **amino acid** if it were **hydrolyzed**. This difference has little effect on the polymer's mechanical or chemical properties, but is sufficient to create a legal distinction.

Properties

- Nylon 6 fibres are tough, possessing high **tensile strength**.
- Fibres have high **elasticity** and lustre.
- Highly resistant to **abrasion**, chemicals like **acids, alkalis**, etc.
- They are wrinkle proof.
- It has the potential to be used as a technical nutrient.
- Nylon 6 absorbs upto 2.4 % of water which results in reduction in tensile strength.
- Nylon 6 is a semicrystalline polyamide.
- Nylon 6 has a much lower melting point than nylon 66. This is a serious disadvantage, as garments made from it must be ironed with considerable care.

Applications

- Bristles for toothbrushes, sutures for surgery, etc.
- Manufacture of hosiery, knitted garments, etc.
- Manufacture of a large variety of threads, ropes, filaments, nets, tire cords, etc.
- Classical guitar strings
- Gun frames, such as those used by Glock, are made with a composite of Nylon 6 and other polymers.

Nylon 6-6

Nylon 6-6, also referred to as nylon 6,6, is a type of **nylon**. Nylon comes in many types, the two most common for textile and plastics industries are: **nylon 6** and nylon 6,6.

Composition

Nylon 6,6 is made of **hexamethylene diamine** and **adipic acid**, which give nylon 6,6 a total of 12 carbon atoms, and its name.

Physical Properties

1. Nylon 6,6 has a melting point of 256°C, high for a synthetic fibre, though not a match for polyesters or aramids such as **Kevlar**. This fact makes it the most resistant to heat and friction and enables it to withstand heatsetting for twist retention.
2. Its long molecular chain results in more sites for hydrogen bonds, creating chemical "springs" and making it very resilient.
3. It has a dense structure with small, evenly spaced pores. This means that nylon 6,6 is difficult to dye, but once dyed it has superior colorfastness and is less susceptible to fading from sunlight and ozone and to yellowing from nitrous oxide.

Manufacture

1. Hexamethylene diamine and adipic acid are combined with water in a reactor. This produces nylon salt. The nylon salt is then sent to an evaporator where excess water is removed.
2. The nylon salt goes into a reaction vessel where a continuous polymerization process takes place. This chemical process makes molten nylon 6,6.
3. The molten nylon 6,6 undergoes a spinning process, where the nylon 6,6 is extruded and sent through a spinnerette, which is a small metal plate with fine holes. The nylon is then air-cooled to form filaments.

4. Nylon's chemical formula is $nHOOC-(CH_2)_4-COOH+n\ H_2N-(CH_2)_6-NH_2 \longrightarrow [-OC-(CH_2)_4-CO-NH-(CH_2)_6-NH-]_n+2nH_2O$ and the part -CO-NH- will stick together becoming Nylon 6,6.

Applications

Carpet fibre

- Apparel
- Airbags
- Tires
- Ropes
- Conveyor Belts
- Hoses

Nylon 6,6's longer molecular chain and denser structure qualifies it as a premium nylon fibre, specified most often by professional architects and designers for use in commercial settings like offices, airports, and other places that get a lot of wear and tear. It is also an excellent choice for residential carpet applications.

Properties of Nylon 66

- Tenacity-elongation at break ranges from 8.8g/d-18% to 4.3 g/d-45%. Its tensile strength is higher than that of wool, silk, rayon, or cotton.
- 100% elastic under 8% of extension
- Specific gravity of 1.14
- Melting point of 263°C
- Extremely chemically stable
- No mildew or bacterial effects
- 4 - 4.5% of moisture regain
- Degraded by light as natural fibres
- Permanent set by heat and steam
- Abrasion resistant
- Lustrous- Nylon fibres have the luster of silk
- Easy to wash
- Can be pre colored or dyed in wide range of colors; dyes are applied to the molten mass of nylon or to the yarn or finished fabric.

- Resilient
- Filament yarn provides smooth, soft, long lasting fabrics
- Spun yarn lend fabrics light weight and warmth

Ballistic Nylon

Ballistic nylon is a thick, tough, synthetic nylon fabric used for a variety of applications. Ballistic nylon was originally developed by the DuPont corporation as a material for flak jackets to be worn by World War II airmen. The term ballistic nylon takes its name from the fact that it was intended to protect its wearers from flying debris and fragmentation caused by bullet or artillery shell impacts. It was not an effective defense against most pistol or rifle rounds. Ballistic nylon was succeeded by Kevlar and other, more effective, bullet resistant fabrics.

Modern applications of ballistic nylon include luggage, cave packs, tool belts, police duty belts, motorcycle jackets, and skin-on-frame kayaks.

Ballistic nylon is of a heavier weight than most nylons used in manufacturing: 18 oz (1050 denier). Denier refers to the weight, not the strength, of the fabric. Laboratory tests have been run and the 1050 denier fabric is the strongest and most durable fabric for its denier weight level. The 1680 denier looks very similar to the 1050 when it is new, but it is actually not as strong as 1050 denier and gets fuzzy as it wears. As it resists abrasion and tearing, ballistic nylon is typically used for luggage and in the bottom of bags made of a lighter nylon or other fabric. Ballistic nylon is hard to dye, so it is often found in black or similar dark colors.

Rip-stop Nylon

Rip-stop nylon is a light-weight nylon fabric with interwoven ripstop reinforcement threads in a crosshatch pattern. It is woven with coarse, strong warp and filling yarns spaced at intervals so that tears will not spread. The same effect can be achieved by weaving two or three of the fine yarns together at intervals.

It is produced in a range of weights and textures, waterproof, water resistant, fire resistant, zero porosity (will not allow air or water through), light, medium and heavy weight. Textures range from soft, silk like to crisp or stiff ones that sound like a paper bag when they are moved.

Rip-stop nylon is often used in yachts for sails and spinnakers hot air balloons, kites, parachutes, camping equipment such as lightweight tents and sleeping bags, flags, banners, and many other applications which require a strong lightweight fabric. Rip-stop reinforcement can also be incorporated into heavier fabrics which require extreme durability, such as those used in the manufacture of luggage and Nomex protective clothing for riot police.

Ejector seat parachutes made with ripstop are woven with an elastic-like fabric so that they stretch to allow more air to pass through them at high speed. Then as the ejector seat slows, the weave closes and acts like a normal parachute. This allows the seat with the pilot in it to slow down gently, otherwise the compression caused could result in injuries to the spine.

It was originally developed as an inexpensive synthetic replacement for silk in the production of parachutes during World War II.

Cordura

CORDURA was the registered name of a certified (nylon) fabric from DuPont [1] . Today, it is the registered name of a high-performance Nylon product manufactured by INVISTA, a wholly-owned division of Koch Industries, Inc. It is used in a wide range of products from luggage and backpacks to boots, military apparel (such as tactical blade sheaths, ammunition pouches, etc.), and performance apparel.

It is designed to be long lasting and resistant to abrasions, tears and scuffs.

As a brand name, CORDURA reaches back to 1929 as a development of Rayon. The product was further developed during World War II and used by the military in tires. From 1966 when new formulations of nylon proved superior, the CORDURA brand was transferred to the nylon product instead.

CORDURA has continued to be developed as a brand by INVISTA with new products applied to an increasing range of items such as:

- luggage
- motorcycle clothing
- anywhere that abrasion resistance is important
- sporting and recreational equipment

Nylon Riots

The **nylon riots** refer to a series of disturbances at American stores created by a nylon stocking shortage.

Nylon was first introduced around 1939 and was rapidly in extremely high demand in the US, with up to 4 million pairs of stockings bought in one day. The riots occurred between August 1945 and March 1946, when Dupont shifted its manufacturing from wartime material to nylon stockings, at the same time launching a promotional campaign. In one of the worst disturbances, in Pittsburgh, 40,000 women queued up for 13,000 pairs of stockings, leading to fights breaking out. It took several months before Dupont was able to ramp up production to meet demand, but until this was able to happen many women went without nylon stockings for months.

6

Acrylic and Modacrylic

Modacylic is a Synthetic co-polymer. Modacrylics are soft, strong, resilient, and dimensionally stable. They can be easily dyed, show good press and shape retention, and are quick to dry. They have outstanding resistance to chemicals and solvents, are not attacked by moths or mildew, and are non-allergenic. Among their uses are in apparel linings, furlike outerwear, paint-roller covers, scatter rugs, carpets, and work clothing and as hair in wigs.

Modacrylic fibres are made from resins that are copolymers *(combinations)* of acrylonitrile and other materials, such as vinyl chloride, vinylidene chloride or vinyl bromide. Modacrylic fibres are either dry spun or wet spun.

Modacrylic Fibre Characteristics

- Soft
- Resilient
- Easy to dye to bright shades

- o Abrasion resistant
- o Flame resistant
- o Quick drying
- o Resistant to acids and alkalies
- o Shape retentive

The low softening temperatures of modacrylic fibres allow them to be stretched, embossed and molded into special shapes. The fibres may be produced with controlled heat shrinkage capacities. When fibres of different shrinkages are mixed in the surface of a pile fabric, the application of heat develops fibres of different lengths, producing a surface that resembles natural fur.

Some Major Modacrylic Fibre Uses

- **Apparel:** Deep-pile coats, trims and linings, simulated fur, wigs and hair pieces, children's sleepwear, career apparel
- **Fabric:** Fleece, knit-pile fabric backings, nonwovens
- **Home Furnishings:** Awnings, blankets, carpets, flame-resistant draperies and curtains, scatter rugs
- **Other Uses:** Filters, industrial fabrics, paint rollers, stuffed toys

General Modacrylic Care Tips — Dry-cleaning or fur-cleaning process is suggested for deep-pile garments. For washable items:

- Machine wash in warm water and add fabric softener during the final rinse cycle.
- If dryer is used, use low setting and remove articles as soon as tumbling cycle has stopped.
- If ironing is required, use low setting. Never use a hot iron. *(For specific instructions, refer to garment's sewn-in care label.)*

Commercial production of modacrylic fibre began in 1949 by Union Carbide Corporation in the United States . Modacrylic and acrylic fibres are similar in composition and at one time were in the same category. In 1960 the Federal Trade Commission decided to separate the two fibres and establish a category for each.

The Federal Trade Commission defines modacrylic fibres as manufactured fibres in which the fibre-forming substance is any long-chain synthetic polymer composed of less than 85%, but at least 35% weight acrylonitrile units except when the polymer qualifies as rubber.

Production

Modacrylic fibres are modified acrylic fibres made from acrylonitriles, but larger amounts of other polymers are added to make the copolymers.

The modacrylic fibres are produced by polymerizing the components, dissolving the copolymer in acetone, pumping the solution into the column of warm air (dry-spun), and stretching while hot.

Modacrylics are creamy or white and are produced in tow and staple form. If looked at in cross section views they have an irregular shape. Modacrylic fibres are also produced in many different lengths, crimp levels, deniers and they can have various shrinkage potentials.

Current modacrylic fibre producers include Solutia Inc. in the U.S. and Kaneka Corporation in Japan.

Properties

A modacrylic has properties that are similar to an acrylic. However, modacrylics are flame retardant and do not combust. The fibres are difficult to ignite and will self-extinguish. In addition to a modacrylic's flame retardant properties it has a relatively high durability that is comparable to wool. Modacrylic fibres have a moderate resistance to abrasion and a very low tenacity.

Modacrylics are poor conductors of heat. The fabrics are soft, warm and resilient but are prone to pilling and matting. Modacrylics display high performance when it comes to appearance retention. The fibres are quite resilient and will not wrinkle. They also have great dimensional stability and high elastic recovery, which gives them the ability to hold their shape.

Care

Modacrylics are sensitive to loss of appearance due to improper care; therefore, it is important to know how to care for modacrylics. Modacrylics are resistant to acids, weak alkalis, and organic solvents. These fibres are also resistant to moths, mildew and sunlight. Modacrylic fabrics can be machine washed using warm water and tumble dried on a low setting. Modacrylic pieces can also be dry-cleaned, however, they should not be steamed and should only be tumbled on cold. Some fabrics may also be cleaned using the furrier method (a special non immersion cleaning process). The fibres are heat sensitive and will shrink at 250 °F and will stiffen at temperatures over 300 °F.

Uses

Modacrylics are used primarily in applications where environmental resistance or flame retardancy is necessary or required. Modacrylics have the ability to combine flame retardancy with a relatively low density, meaning protective gear is not uncomfortably heavy (i.e. shirts and trousers worn by electrical linemen). The combination of flame retardancy and low density is also useful in furnishings, draperies, and outdoor fabrics.

Modacrylics are also commonly used in furlike fabrics, hairpieces, wigs and fleece type fabric. By mixing the various forms of fibres it is easy to create a realistic fur look. The fabrics can then be sheared or embossed to create a closer resemblance to fur. The heat-sensitivity of modacrylic also allows wigs and hairpiece to be curled and heat styled without damage.

Modacrylics are also used in fleece, knit-pile fabric backings, and nonwoven fabrics. Other uses of modacrylics include paint rollers, industrial fabrics, stuffed toys and filters.

Modacrylic is a manufactured fibre in which the fibre-forming substance is any long-chain synthetic polymer composed of less than 85 percent but at least 35 percent by weight of acrylonitrile units. As the definition, indicates, variations are possible in producing this type of fibre. Modacrylics are made from resins that are copolymers or combinations of acrylonitrile and other compounds, such as vinyl chloride, vinylidene chloride, or vinylidene dicyanide. These chemicals are made of such elements as carbon, hydrogen, nitrogen, and chlorine, which are obtained from natural gas, coal, air, salt, and water. Trademarks of two modacrylics are Verel and SEF, which, in turn, have modifications. These modacrylic fibres have similar properties to those of acrylics and are flame-resistant.

Acrylic

Acrylic is a useful, clear plastic that resembles glass, but has properties that make it superior to glass in many ways. Common brands of high-grade Acrylic include Polycast, Lucite and Plexiglass.

There are two basic types of Acrylic: extruded and cell cast. Extruded or "continuous cast" Acrylic is made by a less expensive process, is softer, can scratch easier and may contain impurities. Cell cast Acrylic is a higher quality Acrylic and U.S. domestic cell cast is a good choice for applications that require the best. Imported cell cast Acrylic is often manufactured to lesser standards.

Acrylic is used to make various products, such as shower doors, bath enclosures, windows and skylights. It is chosen over glass for many reasons. It is many times stronger than glass, making it much more impact resistant and therefore safer. Falling against an Acrylic shower door will not likely break it. Baseballs that crash through glass windows will, in most cases, bounce off Acrylic windows. Acrylic also insulates better than glass, potentially saving on heating bills.

Another great advantage of Acrylic is that it is only half as heavy as glass. This makes working with Acrylic much easier. It can also be sawed, whereas glass must be scored.

Adding to this favorable array of properties, a transparency rate of 93% makes Acrylic the clearest material known. Very thick glass will have a green tint, while Acrylic remains clear.

A unique property of plastic is its ability to be shaped. Bow-front aquariums are beautiful examples of Acrylic's wonderful properties. There are also no seams in Acrylic structures, as chemical welding at the molecular level actually "melts" seams into one piece of solid material. Seams that are welded and polished are invisible.

There are some misconceptions about Acrylic, namely that it yellows, turns brittle and cracks over time. Though this might be true of very cheap forms of plastic, it is not so with Acrylic. For example, the fighter planes of WWII have Acrylic bubble-tops. Airplane windows are also Acrylic. If taken care of, Acrylic remains new looking regardless of age or exposure to sun. Some people worry that Acrylic scratches too easily, but unlike glass, scratches can be easily buffed out of Acrylic.

For all of its advantages, there are two disadvantages of Acrylic: it is more expensive than glass, and if exposed to a direct flame it will melt and eventually burn.

Today Acrylic is used more than ever. Virtually all major public aquariums now build display tanks out of Acrylic. You will also find Acrylic in malls, institutions, prisons, hospitals and commercial buildings. Acrylic just over one inch thick

(32mm) is bullet resistant. The Presidential motorcade, the Pope's booth-vehicle, teller enclosures and drive-through window enclosures all feature bullet-resistant Acrylic.

If upgrading the windows in your house, remodeling your bathroom, or adding a beautiful aquarium, consider Acrylic. It may cost a little more than glass, but its sheer clarity, light weight and insulating properties make it a superior choice for many applications.

Acrylic fibres are synthetic fibres made from a polymer (Polyacrylonitrile) with an average molecular weight of ~100,000, about 1900 monomer units. To be called acrylic in the U.S, the polymer must contain at least 85% acrylonitrile monomer. Typical comonomers are vinyl acetate or methyl acrylate. The Dupont Corporation created the first acrylic fibres in 1941 and trademarked them under the name "Orlon".

Production

The polymer is formed by free-radical polymerization in aqueous suspension. The fibre is produced by dissolving the polymer in a solvent such as N,N-dimethylformamide or aqueous sodium thiocyanate, metering it through a multi-hole spinnerette and coagulating the resultant filaments in an aqueous solution of the same solvent (wet spinning) or evaporating the solvent in a stream of heated inert gas (dry spinning). Washing, stretching, drying and crimping complete the processing. Acrylic fibres are produced in a range of deniers, typically from 1 to 15 as cut staple or as a 500,000 to 1 million filament tow. End uses include sweaters, hats, hand-knitting yarns, rugs, awnings, boat covers, and upholstery; the fibre is also used as a precursor for carbon fibre. Production of acrylic fibres is centered in the Far East, declining in Europe and now shut down (except for precursor) in the U.S. Former U.S. brands of acrylic were Acrilan (Monsanto), Creslan (American Cyanamid), and Orlon (DuPont).

Textile Uses

Acrylic is lightweight, soft, and warm, with a wool-like feel. It dyes very well and has excellent colorfastness. It is resilient, retains its shape, and resists shrinkage and wrinkles. Acrylic has recently been used in clothing as a less expensive alternative to cashmere, due to the similar feeling of the materials. The disadvantages of acrylic are that it tends to fuzz (or pill) easily and that it does not insulate the wearer as well as cashmere. Many products like fake pashmina or cashmina use this fibre to create the illusion of cashmere to the consumer.

Acrylic is resistant to moths, oils, and chemicals, and is very resistant to deterioration from sunlight exposure. However, static and pilling can be a problem. Acrylic can irritate the skin of people with eczema.

History of Acrylic

Acrylic plastic refers to a family of synthetic, or man-made, plastic materials containing one or more derivatives of acrylic acid. The most common acrylic plastic is polymethyl methacrylate (PMMA), which is sold under the brand names of Plexiglas, Lucite, Perspex, and Crystallite. PMMA is a tough, highly transparent material with excellent resistance to ultraviolet radiation and weathering. It can be colored, molded, cut, drilled, and formed. These properties make it ideal for many applications including airplane windshields, skylights, automobile taillights, and outdoor signs. One notable application is the ceiling of the Houston Astrodome which is composed of hundreds of double-insulating panels of PMMA acrylic plastic.

Like all plastics, acrylic plastics are polymers. The word polymer comes from the Greek words *poly*, meaning many, and *meros*, meaning a part. A polymer, therefore, is a material made up of many molecules, or parts, linked together like a chain. Polymers may have hundreds, or even thousands, of molecules linked together. More importantly, a polymer is a material that has properties entirely different than its component parts. The process of making a polymer, known as polymerization, has been likened to shoveling scrap glass, copper, and other materials into a box, shaking the box, and coming back in an hour to find a working color television set. The glass, copper, and other component parts are still there, but they have been reassembled into something that looks and functions entirely differently.

The first plastic polymer, celluloid, a combination of cellulose nitrate and camphor, was developed in 1869. It was based on the natural polymer cellulose, which is present in plants. Celluloid was used to make many items including photographic film, combs, and men's shirt collars.

In 1909, Leo Baekeland developed the first commercially successful synthetic plastic polymer when he patented phenol formalde-hyde resin, which he named Bakelite. Bakelite was an immediate success. It could be machined and molded. It was an excellent electrical insulator and was resistant to heat, acids, and weather. It could also be colored and dyed for use in decorative objects. Bakelite plastic was used in radio, telephone, and electrical equipment, as well as counter tops, buttons, and knife handles.

Acrylic acid was first prepared in 1843. Methacrylic acid, which is a derivative of acrylic acid, was formulated in 1865. When methacrylic acid is reacted with methyl alcohol, it results in an ester known as methyl methacrylate. The polymerization process to turn methyl methacrylate into polymethyl methacrylate was discovered by the German chemists Fittig and Paul in 1877, but it wasn't until 1936 that the process was used to produce sheets of acrylic safety glass commercially. During World War II, acrylic glass was used for periscope ports on submarines and for windshields, canopies, and gun turrets on airplanes.

Raw Materials

Methyl methacrylate is the basic molecule, or monomer, from which polymethyl methacrylate and many other acrylic plastic polymers are formed. The chemical notation for this material is $CH_2=C(CH_3)\ COOCH_3$. It is written in this format, rather than the more common chemical notation $C_5H_8O_2$, to show the double bond (=) between the two carbon atoms in the middle. During polymerization, one leg of this double bond breaks and links up with the middle carbon atom of another methyl methacrylate molecule to start a chain. This process repeats itself until the final polymer is formed.

Methyl methacrylate may be formed in several ways. One common way is to react acetone [CH_3COCH_3] with sodium cyanide [NaCN] to produce acetone cyanhydrin [$(CH_3)_2C(OH)CN$]. This in turn is reacted with methyl alcohol [CH_3OH] to produce methyl methacrylate.

Other similar monomers such as methyl acrylate [CH_2=CHCOOCH,] and acrylonitrile [$CH_2=CHCN$] can be joined with methyl methacrylate to form different acrylic plastics. When two or more monomers are joined together, the result is known as a copolymer. Just as with methyl methacrylate, both of these monomers have a double bond on the middle carbon atoms that splits during polymerization to link with the carbon atoms of other molecules. Controlling the proportion of these other monomers produces changes in elasticity and other properties in the resulting plastic.

The Manufacturing Process

Acrylic plastic polymers are formed by reacting a monomer, such as methyl methacrylate, with a catalyst. A typical catalyst would be an organic peroxide. The catalyst starts the reaction and enters into it to keep it going, but does not become part of the resulting polymer.

Acrylic plastics are available in three forms: flat sheets, elongated shapes (rods and tubes), and molding powder. Molding powders are sometimes made by a process known as suspension polymerization in which the reaction takes place between tiny droplets of the monomer suspended in a solution of water and catalyst. This results in grains of polymer with tightly controlled molecular weight suitable for molding or extrusion.

Acrylic plastic sheets are formed by a process known as bulk polymerization. In this process, the monomer and catalyst are poured into a mold where the reaction takes place. Two methods of bulk polymerization may be used: batch cell or continuous. Batch cell is the most common because it is simple and is easily adapted for making acrylic sheets in thicknesses from 0.06 to 6.0 inches (0.16-15 cm) and widths from 3 feet (0.9 m) up to several hundred feet. The batch cell method may also be used to form rods and tubes. The continuous method is quicker and involves less labor. It is used to make sheets of thinner thicknesses and smaller widths than those produced by the batch cell method.

We will describe both the batch cell and continuous bulk polymerization processes typically used to produce transparent polymethyl methacrylic (PMMA) sheets.

Batch Cell Bulk Polymerization

1. The mold for producing sheets is assembled from two plates of polished glass separated by a flexible "window-frame" spacer. The spacer sits along the outer perimeter of the surface of the glass plates and forms a sealed cavity between the plates. The fact that the spacer is flexible allows the mold cavity to shrink during the polymerization process to compensate for the volume contraction of the material as the reaction goes from individual molecules to linked polymers. In some production applications, polished metal plates are used instead of glass. Several plates may be stacked on top of

each other with the upper surface of one plate becoming the bottom surface of the next higher mold cavity. The plates and spacers are clamped together with spring clamps.

2. An open comer of each mold cavity is filled with a pre-measured liquid syrup of methyl methacrylate monomer and catalyst. In some cases, a methyl methacrylate prepolymer is also added. A prepolymer is a material with partially formed polymer chains used to further help the polymerization process. The liquid syrup flows throughout the mold cavity to fill it.

3. The mold is then sealed and heat may be applied to help the catalyst start the reaction.

4. As the reaction proceeds, it may generate significant heat by itself. This heat is fanned off in air ovens or by placing the molds in a water bath. A programmed temperature cycle is followed to ensure proper cure time without additional vaporization of the monomer solution. This also prevents bubbles from forming. Thinner sheets may cure in 10 to 12 hours, but thicker sheets may require several days.

5. When the plastic is cured, the molds are cooled and opened. The glass or metal plates are cleaned and reassembled for the next batch.

6. The plastic sheets are either used as is or are annealed by heating them to 284-302°F (140-150°C) for several hours to reduce any residual stresses in the material that might cause warping or other dimensional instabilities.

7. Any excess material, or flash, is trimmed off the edges, and masking paper or plastic film is applied to the surface of the finished sheets for protection during handling and shipping. The paper or film is often marked with the material's brand name, size, and handling instructions. Conformance with applicable safety or building code standards is also noted.

Continuous Bulk Polymerization

1. The continuous process is similar to the batch cell process, but because the sheets are thinner and smaller, the

process times are much shorter. The syrup of monomer and catalyst is introduced at one end of a set of horizontal stainless steel belts running parallel, one above the other. The distance between the belts determines the thickness of the sheet to be formed.

2. The belts hold the reacting monomer and catalyst syrup between them and move it through a series of heating and cooling zones according to a programmed temperature cycle to cure the material.

3. Electric heaters or hot air may then anneal the material as it comes out of the end of the belts.

4. The sheets are cut to size and masking paper or plastic film is applied.

Quality Control

The storage, handling, and processing of the chemicals that make acrylic plastics are done under controlled environmental conditions to prevent contamination of the material or unsafe chemical reactions. The control of temperature is especially critical to the polymerization process. Even the initial temperatures of the monomer and catalyst are controlled before they are introduced into the mold. During the entire process, the temperature of the reacting material is monitored and controlled to ensure the heating and cooling cycles are the proper temperature and duration.

Samples of finished acrylic materials are also given periodic laboratory analysis to confirm physical, optical, and chemical properties.

Toxic Materials, Safety Considerations, and Recycling

Acrylic plastics manufacturing involves highly toxic substances which require careful storage, handling, and disposal. The polymerization process can result in an explosion if not monitored properly. It also produces toxic fumes. Recent legislation requires that the polymerization process be carried out in a closed environment and that the fumes be cleaned, captured, or otherwise neutralized before discharge to the atmosphere.

Acrylic plastic is not easily recycled. It is considered a group 7 plastic among recycled plastics and is not collected for recycling in most communities. Large pieces can be reformed into other useful objects if they have not suffered too much stress, crazing, or cracking, but this accounts for only a very small portion of the acrylic plastic waste. In a landfill, acrylic plastics, like many other plastics, are not readily biodegradable. Some acrylic plastics are highly flammable and must be protected from sources of combustion.

The Future of Acrylic

The average annual increase in the rate of consumption of acrylic plastics has been about 10%. A future annual growth rate of about 5% is predicted. Despite the fact that acrylic plastics are one of the oldest plastic materials in use today, they still hold the same advantages of optical clarity and resistance to the outdoor environment that make them the material of choice for many applications.

Acrylic Glass

Polymethyl methacrylate) (**PMMA**) polymethyl 2-methylpropenoate) is a thermoplastic and transparent plastic. Chemically, it is the synthetic polymer of methyl methacrylate. It is sold by the trade names **Plexiglas, Vitroflex, Limacryl, R-Cast, Per-Clax, Perspex, Plazcryl, Acrylex, Acrylite, Acrylplast, Altuglas, Polycast, Oroglass, Optix** and **Lucite** and is commonly called **acrylic glass**, simply **acrylic, perspex** or **plexiglas**. Acrylic, or acrylic fibre, can also refer to polymers or copolymers containing polyacrylonitrile. The material was developed in 1928 in various laboratories and was brought to market in 1933 by Rohm and Haas Company.

PMMA is often used as an alternative to glass, and in competition with polycarbonate (PC). It is often preferred because of its moderate properties, easy handling and processing, and low cost, but behaves in a brittle manner when loaded, especially under an impact force. To produce 1 kg of PMMA, about 2 kg of petroleum is needed. PMMA ignites at 460 °C and burns completely to form only carbon dioxide and water.[

History

The first acrylic acid was created in 1843. Methacrylic acid, derived from acrylic acid, was formulated in 1865. The reaction between methacrylic acid and methyl alcohol results in the ester methyl methacrylate. The German chemists Fittig and Paul discovered in 1877 the polymerization process that turns methyl methacrylate into polymethyl methacrylate. In 1933 the German chemist Otto Röhm patented and registered the brand name PLEXIGLAS. In 1936 the first commercially viable production of acrylic safety glass began. During World War II acrylic glass was used for submarine periscopes, and windshields, canopies, and gun turrets for airplanes. [1]

Synthesis

PMMA is routinely produced by emulsion polymerization, solution polymerization and bulk polymerization. Generally radical initiation is used (including living polymerization methods), but anionic polymerization of PMMA can also be performed.

Processing

Thermoplastic PMMA is typically processed at 240–250 °C. All common molding processes may be used, including injection molding, compression molding and extrusion. The highest quality PMMA sheets are produced by cell casting, but in this case, the polymerization and molding steps occur concurrently. The strength of the material is higher than molding grades owing to its extremely high molecular mass. Rubber toughening has been used to increase the strength of

PMMA owing to its brittle behavior in response to applied loads.

PMMA can be joined using cyanoacrylate cement (so-called "Superglue"), with heat (melting), or by using solvents such as di- or trichloromethane to dissolve the plastic at the joint which then fuses and sets, forming an almost invisible weld.

Scratches may easily be removed by polishing or by heating the surface of the material.

Laser cutting may be used to form intricate designs from PMMA sheets. PMMA vaporises to gaseous compounds (including its monomers) upon laser cutting, so a very clean cut is made, and cutting is performed very easily. In this respect PMMA has an advantage over competing polymers such as polystyrene and polycarbonate, which require higher laser powers and give more messy and charred laser cuts.

Properties

PMMA:

- has a density of 1,150–1,190 kg/m^3. This is less than half the density of glass, and similar to that of other plastics.
- has a good impact strength higher than that of glass or polystyrene, but significantly lower than that of polycarbonate or engineering polymers. In the majority of applications, it will not shatter but instead breaks into large dull pieces.
- is softer and more easily scratched than glass. Scratch-resistant coatings (which may also have other functions) are often added to PMMA sheets.
- transmits up to 98% of visible light (3 mm thickness), and gives a reflection of about 4% from each of its surfaces on account of its refractive index of 1.4893 to 1.4899.
- filters ultraviolet (UV) light at wavelengths below about 300 nm. Some manufacturers add coatings or additives to PMMA to improve absorption in the 300–400 nm range.

- allows infrared light of up to 2800 nm wavelength to pass. IR of longer wavelengths, up to 25,000 nm, are essentially blocked. Special formulations of colored PMMA exist to allow specific IR wavelengths to pass while blocking visible light (for remote control or heat sensor applications, for example).
- has excellent environmental stability compared to other plastics such as polycarbonate, and is therefore often the material of choice for outdoors applications.
- has poor resistance to solvents, as it swells and dissolves easily. It also has poor resistance to many other chemicals on account of its easily hydrolyzed ester groups.

Modification of Properties

Pure polymethyl methacrylate) homopolymer is rarely sold as an end product, since it is not optimized for most applications. Rather, modified formulations with varying amounts of other comonomers, additives, and fillers are created for uses where specific properties are required. For example,

- A small amount of acrylate comonomers are routinely used in PMMA grades destined for heat-processing, since this stabilizes the polymer to depolymerization ("unzipping") during processing.
- Comonomers such as butyl acrylate are often added to improve impact strength.
- Comonomers such as methacrylic acid can be added to increase the glass transition temperature of the polymer for higher temperature use such as in lighting applications.
- Plasticizers may be added to improve processing properties, lower the glass transition temperature, or improve impact properties.
- Dyes may be added to give color for decorative applications, or to protect against (or filter) UV light.
- Fillers may be added to improve cost-effectiveness.

Related Polymer Polymethyl Acrylate)

The polymer of methyl acrylate, PMA or polymethyl acrylate), is similar to polymethyl methacrylate), except for the lack of methyl groups on the backbone carbon chain. PMA is a soft white rubbery material that is softer than PMMA because its long polymer chains are thinner and smoother and can more easily slide past each other.

Uses

PMMA or Acrylic is a versatile material and has been used in a wide range of fields and applications.

Impact resistant substitute for glass

- PMMA Acrylic glass is commonly used for constructing residential and commercial aquariums. Designers started building big aquariums when polymethyl methacrylate) could been used. The spectacular size of both flat panels and tunnels in aquariums such Monetary Bay, Tokyo Sea Life Park, Osaka and Nagoya Aquariums would have been possible with the introduction of acrylic.
- PMMA is used in the lenses of exterior lights of automobiles.
- The spectator protection in ice hockey stadiums is made from PMMA.
- Motorcycle helmet visors
- Police vehicles for riot control often have the regular glass replaced with acrylic to protect the occupants from thrown objects.
- Acrylic is used for viewing ports and even complete hulls of submersibles, such as the Alicia submarine's viewing spheres and the Bathyscaphe Trieste's windows.
- Polycast acrylic sheet is the most widely used material in aircraft transparencies (windows). In applications where the aircraft is pressurized, stretched acrylic is used.
- Acrylic is an important material in the making of certain lighthouse lenses.

Daylight Redirection

- Laser cut acrylic panels have been used to redirect sunlight into a light pipe and, from there, to spread it into a room. Their developers Veronica Garcia Hansen, Ken Yeang, and Ian Edmonds were awarded the Far East Economic Review Innovation Award in bronze for this technology in 2003.[7][8]
- Attenuation being quite strong for distances over one meter (more than 90% intensity loss for a 3000 K source), acrylic broadband light guides are then dedicated mostly to decorative uses.
- Pairs of acrylic sheets with a layer of microreplicated prisms between the sheets can have reflective and refractive properties that let them redirect part of incoming sunlight in dependence on its angle of incidence. Such panels act as miniature light shelves. Such panels have been commercialized for purposes of daylighting, to be used as a window or a canopy such that sunlight descending from the sky is directed to the ceiling or into the room rather than to the floor. This can lead to a higher illumination of the back part of a room, in particular when combined with a white ceiling, while having a slight impact on the view to the outside compared to normal glazing.

Medical Technologies and Implants

- PMMA has a good degree of compatibility with human tissue, and can be used for replacement intraocular lenses in the eye when the original lens has been removed in the treatment of cataracts. Historically, hard contact lenses were frequently made of this material. Soft contact lenses are often made of a related polymer, where acrylate monomers containing one or more hydroxyl groups make them hydrophilic.
- In orthopaedics, PMMA bone cement is used to affix implants and to remodel lost bone. It is supplied as a powder with liquid methyl methacrylate (MMA). When mixed these

yield a dough-like cement that gradually hardens. Surgeons can judge the curing of the PMMA bone cement by pressing their thumb on it. Although PMMA is biologically compatible, MMA is considered to be an irritant and a possible carcinogen. PMMA has also been linked to cardiopulmonary events in the operating room due to hypotension. [12] Bone cement acts like a grout and not so much like a glue in arthroplasty. Although sticky, it does not bond to either the bone or the implant, it primarily fills the spaces between the prosthesis and the bone preventing motion. A big disadvantage to this bone cement is that it heats to quite a high temperature while setting and because of this it kills the bone in the surrounding area. It has a Young's modulus between cancellous bone and cortical bone. Thus it is a load sharing entity in the body not causing bone resorption. [13]

- Dentures are often made of PMMA, and can be colour-matched to the patient's teeth & gum tissue. In cosmetic surgery, tiny PMMA micro spheres suspended in some biological fluid are injected under the skin to reduce wrinkles or scars permanently.

Artistic and Aesthetic Uses

- Acrylic paint essentially consists of PMMA suspended in water; however since PMMA is hydrophobic, a substance with both hydrophobic and hydrophilic groups needs to be added to facilitate the suspension.
- Modern furniture makers, especially in the 1960s and 1970s, seeking to give their products a space age aesthetic incorporated Lucite and other PMMA products into their designs, especially office chairs. Many other products (for example, guitars) are sometimes made with acrylic glass to make the commonly opaque objects translucent.
- Perspex has been used as a surface to paint on, for example by Salvador Dalí.

- Diasec is a process which uses acrylic glass as a substitute for normal glass in picture framing. This is done for its relatively inexpensive cost, light weight, shatter-resistant nature, aesthetic reasons and for the fact that it can be ordered in larger sizes than standard picture-framing glass.
- From approximately the 1960s onward, sculptors and glass artists began using acrylics, especially taking advantage of the material's flexibility, light weight, cost and its capacity to refract and filter light.
- Sometimes used to make a Deal toy (in the world of finance or investment banking).
- Used in contact juggling.

Other Uses

- PMMA is used as a shield to stop beta radiation emitted from radioisotopes.
- PMMA was used in laserdisc optical media. (CDs and DVDs use the more expensive polycarbonate for higher impact resistance.)
- PMMA-based optical media is under development for the "TeraDisc" next-generation 3D optical data storage solution by Mempile.
- Artificial fingernails are made of acrylic.
- In the 1960s, luthier Dan Armstrong developed a line of electric guitars and basses whose bodies were made completely of acrylic. These instruments were marketed under the Ampeg brand. Ibanez and BC Rich have also made acrylic guitars.
- Recently a blacklight-reactive tattoo ink using PMMA microcapsules was developed. This ink is reportedly safe for use, and claims to be Food and Drug Administration (FDA) approved for use on wildlife that may enter the food supply.
- In semiconductor research and industry, PMMA aids as a resist in the electron beam lithography process. A

solution consisting of the polymer in a solvent is used to spin coat silicon and other semiconducting and semi-insulating wafers with a thin film. Patterns on this can be made by an electron beam (using an electron microscope), deep UV light (shorter wavelength than the standard photolithography process), or X-rays. Exposure to these creates chain scission or (de-cross-linking) within the PMMA, allowing for the selective removal of exposed areas by a chemical developer, making it a positive photoresist. PMMA's advantage is that it allows for extremely high resolution (nanoscale) patterns to be made. It is an invaluable tool in nano technology.

- Small strips of PMMA are used as dosimeter devices during the Gamma Irradiation process. The optical density of PMMA changes as the Gamma dose increases and can be measured with a spectrophotometer.
- It is used as a light guide for the backlights in TFT-LCDs.
- Plastic optical fibre used for short distance communication is made from PMMA, and perfluorinated PMMA, clad with fluorinated PMMA, in situations where its flexibility and cheaper installation costs outweigh its poor heat tolerance and higher attenuation over glass fibre.
- Sheets of PMMA are commonly used in the sign industry to make flat cut out letters in thicknesses typically varying from 0.125" to 1". These letters may be used alone to represent a company's name and/or logo, or they may be a component of channel letters which are neon or LED illuminated sign "can" commonly seen worldwide. Acrylic's attractiveness, durability and resistance to warping makes it an ideal interior and exterior sign material.
- Ludwig-Musser makes a line of acrylic drums called Vistalites. They are well known as being used by Led Zeppelin drummer John Bonham.

- Ear stretching jewelry is also commonly formed out of PMMA, due to its inert nature and shatter-proof qualities. PMMA jewelry is fade-proof, odorless, durable, and easy to polish, giving it that modern stylish sheen.
- The case of Apple Computer's Power Mac G4 Cube was made out of PMMA.
- WWE's Elimination Chamber match has 4 "Containment Pods," each with a metal outlining and PMMA walls.
- Acrylic is also used extensively throughout the sign industry as a component of wall signs where it may be a backplate, painted on the surface or the backside, a faceplate with additional raised lettering or even photographic images printed directly to it, or a spacer to separate sign components. One of the most popular sheets is a non-glare, translucent which is sold in 1/16 or 1/8 in thicknesses.
- Acrylic is often used in the manufacturing of bongs used for smoking tobacco, herbs, or cannabis.
- Acrylic can be made into a cube to preserve a sport's championship ring for display, (some owners have the ring removed).
- Lucite was used in the late 40's to mid 50's in automobile hood ornaments, most typically used in Pontiac hood ornaments.

General properties of acrylic sheet:

- Strong and resists weathering
- Flexible when compared with glass
- Less breakable than glass
- Abrasion resistant
- Can withstand sunlight for long durations
- Resistant to most chemicals and industrial fumes
- Can transmit or filter ultraviolet light
- Can be cleaned easily
- Can be cut by various methods
- Corrosion resistant

- Good insulator.

Other general properties include the ability to transmit and control light. They are also stable against discoloration, and have superior dimensional stability, as you would notice if you have an acrylic brochure display, counter display, donation / suggestion box or point of purchase (p.o.p) display. Possessing an excellent combination of structural and thermal properties, clear acrylic plastic is as transparent as the finest optical glass:

- Possess a light transmittance of 92%
- Low haze level of approximately 1%
- Index of refraction of 1.49

It also has the ability to be injected with color, producing a full spectrum of transparent, translucent, or opaque colors depending on your needs. This process does no harm in terms of long-term durability; colored acrylics can be used outdoors for a long time. Why? They are formulated to filter ultraviolet energy in the 360-nm and lower band. Other acrylic formulations are opaque to UV light or provide reduced UV transmission.

And how about mechanical properties? Although not known for having many, acrylics can be used for short-term loading. If the intended use is long-term, stresses must be limited to 1,500 psi to avoid surface cracking and deterioration.

Acrylic does well in the cold, as the impact resistance of standard formulations is maintained in these conditions. It should be noted that high-impact acrylic grades have greater impact strength than standard grades at room temperature, but impact strength decreases as temperature drops. Some types of acrylic are even known to resist bullets!

Acrylic plastics are highly scratch resistant, especially among other thermoplastics. It's a good idea, however, to ensure proper maintenance and cleaning. Keep in mind that abrasion-resistant acrylic sheet is available and has the same optical and impact properties as standard grades. You will see this in many of our brochure displays, racks, counter displays, donation / suggestion boxes and point of purchase (p.o.p) displays.

This brings up a good point - its versatility and adaptability. Are you aware that Jet-aircraft cabin windows are made from acrylic sheet? They do this by inducing molecular orientation during forming. This proves the potential strength of acrylic sheet. How do they react to other chemicals and compounds? Acrylic sheet and moldings resist solutions of inorganic acids and alkalies and aliphatic hydrocarbons such as VM&P naphtha, as well as most detergent solutions and cleaning agents. They are attacked by chlorinated and aromatic hydrocarbons, esters, and ketones.

Acrylic sheets are extremely durable and have weather resistant properties. They can be cleaned very easily and thus are excellent for making poster frames and sign holders

Acrylic is a rigid plastic with a high degree of transparency. It is resistant to inorganic acids and alkalis but is attacked by a wide range of organic solvents. Acrylic's clarity and stability make it very suitable for the manufacture of burets and in sheet form it may be cemented to produce tanks, trays, racks etc. Acrylic may be used at temperatures up to 70°C continuously and to 90°C for short periods. 10mm thick Acrylic provides an effective barrier for Beta Radiation Protection.

Acrylic Resistance:

- Excellent resistance (no attack) to Mineral Oils
- Good resistance (minor attack) to Dilute Acids, Aldehydes and Aliphatic Hydrocarbons
- Limited resistance (moderate attack and suitable for short term use only) to Bases
- Poor resistance Poor resistance (not recommended for use) with Concentrated Acids, Alcohols, Esters, Aromatic

and Halogenated Hydrocarbons, Ketones, Vegetable Oils and Oxidizing Agents.

Acrylic Quick Facts:

- Maximum Temperature: 194°F 90°C
- Minimum Temperature: -76°F -60°C
- Autoclavable: No
- Melting Point: 212°F 100°C
- Tensile Strength: 7,000 psi
- Hardness: R120
- UV Resistance: Good
- Clear
- Rigid
- Specific Gravity: 1.18

Acrylic Fabrication:

- Nearly crystal clear, Acrylic is an excellent choice for display cases, vitrines, and point of purchase items
- Thicker Acrylic material can shield against beta radiation
- Available in a wide array of forms and colors
- Machineable and heat-bendable
- Acrylic can be solvent bonded
- Extruded acrylic can be flame polished to obtain a glass-like finish
- Applications include aquariums, picture frames, shelves, and cabinets

Acrylic Thermoplastic Material

Acrylic is the common term for the synthetic polymer, polymethyl methacrylate (PMMA). Since it's inception to the market, acrylic has significantly increased in both use and benefit. This transparent thermoplastic material is now sold under various trade names including but not limited to Plexiglass, Perspex, Acrylite, Acrylplast, and Lucite.

Methyl methacrylate is the basic monomer from which acrylic is produced and can be formed in several different ways depending on the chemical reaction of the acrylic formulation. Utilizing different methods and controlling elasticity proportions can adjust the properties of acrylic for compliance with most application requirements. Aline maintains close working relationships with referred acrylic suppliers to obtain the precise formulation that will result in the highest level of performance for your product or system.

Acrylic Plastic Properties

When heated to 100°C, acrylic can be easily formed and manufactured in sheets, rods, and tubes or injection molded as powder, cast, or sheets. Colorless and highly transparent, acrylic resins will not discolor or fade when exposed to light over time. Acrylic plastic can transmit both ultraviolet and white light, which makes it highly resistant to ultraviolet radiation and weathering. Although acrylic is one of the few plastics that cannot be easily recycled, it can be reused if manufactured as a large piece and does not show signs of crazing or cracking.

Advantages of Acrylic Plastic Include:

- Ease of plastic injection molding and manufacturing
- Lightweight
- Impact resistant
- Weather and heat resistant
- Tough, durable
- Dimensional stability
- Excellent clarity

A tough plastic, acrylic does not shatter even under high impact. Its resistance to oxidizing agents enables acrylic to perform well in water. With a tensile strength of 8,000 pounds per square inch and a compressive and flexural strength of 15,000 pounds per square inch, acrylic can be used as a great substitute for glass. Rigidity, durability, and stability all make acrylic an excellent choice for a variety of applications, but perhaps its most distinct feature is its optical clarity.

Although acrylic is inherently more scratch resistant than other thermoplastic materials, it is softer and more easily scratched than glass. Also, acrylic plastic has a low resistance to abrasion and subsequently is subjected to crazing when used in replace of glass. Aline can help avoid potential scratches and abrasion by coating the surface for superior finish qualities. Whether you need a custom plastic part or component for outdoor applications where transparent materials are required or you need a lightweight material for applications where glass was traditionally used, custom plastic injection molded and manufactured acrylic plastic for Aline Components provides a long, high performance service life.

Acrylic Fibre

Acrylic fibres are produced from acrylonitrile, a petrochemical. The acrylonitrile is usually combined with small amounts of other chemicals to improve the ability of the resulting fibre to absorb dyes. Some acrylic fibres are dry spun and others are wet spun. Acrylic fibres are used in staple or tow form. For a detailed production flowchart (wet and dry spun), go here.

Acrylic fibres are modified to give special properties best suited for particular end-uses. They are unique among synthetic fibres because they have an uneven surface, even when extruded from a round-hole spinneret.

Acrylic Fibre Characteristics

- Outstanding wickability & quick drying to move moisture from body surface
- Flexible aesthetics for wool-like, cotton-like, or blended appearance
- Easily washed, retains shape
- Resistant to moths, oil, and chemicals
- Dyeable to bright shades with excellent fastness
- Superior resistance to sunlight degradation

Some Major Acrylic Fibre Uses

- **Apparel:** Sweaters, socks, fleece wear, circular knit apparel, sportswear and childrens wear
- **Home Furnishings:** Blankets, area rugs, upholstery, pile; luggage, awnings, outdoor furniture
- **Other Uses:** Craft yarns, sail cover cloth, wipe cloths
- **Industrial Uses:** Asbestos replacement; concrete and stucco reinforcement

Federal Trade Commission Definition for Acrylic Fibre: A manufactured fibre in which the fibre forming substance is any long-chain synthetic polymer composed of at least 85% by weight of acrylonitrile units $(-CH_2-CH[CN]-)_x$.

General Acrylic Fibre Care Tips

- Wash delicate items by hand in warm water. Static electricity may be reduced by using a fabric softener in every third or fourth washing. Gently squeeze out water, smooth or shake out garment and let dry on a non-rust hanger. *(Sweaters, however, should be dried flat.)*
- When machine washing, use warm water and add a fabric softener during the final rinse cycle.
- Machine dry at a low temperature setting. Remove garments from dryer as soon as tumbling cycle is completed.
- If ironing is required, use a moderately warm iron. *(For specific instructions, refer to garment's sewn-in care label.)*

7

Introduction of Olefin

Olefin is a manufactured fibre in which the fibre-forming substance is any long-chain synthetic polymer composed of at least 85 percent by weight of ethylene, propylene, or other olefin units. As the definition indicates, variations are possible in producing this type of fibre. The carbon-hydrogen chains polymerized to produce this fibre are obtained from the cracking of petroleum. Polyethylene, the original olefin fibre, has been superseded by polypropylene. The fibre is produced by a number of manufacturers under their respective trademarks. Polypropylene fibre is thermoplastic, resilient, lightweight, and resistant to mildew and many chemicals. It is used for apparel and carpets, particularly for out-door use, as well as industrial applications.

Olefin, also known as polypropylene or polyethylene, is a long-chain polymer synthetic fibre. It is created when ethylene and/or propylene gases are polymerized under very specific conditions. Olefin was first manufactured in the late 1950s, and has a myriad of applications in manufacturing, household products and clothing. Use of Olefin has steadily increased since the 1960s, and continues to gain in popularity every year as new uses for this easy, durable material are found.

Olefin is created when the polymers are melted to a liquid, then run through a machine called a *spinneret*, which forces the product into a long fibre through small holes. This fibre is then used to make the fabric, or the end product in which it will be used. Because Olefin resists dyeing, any dye must be added during this melting process, instead of to the final fabric or product. It is highly favored due to its easy, inexpensive manufacturing process. Surprisingly, Olefin is relatively environmentally "friendly" due to the few byproducts produced during manufacturing. Olefin is also easily recycled.

There are several benefits of using Olefin. It is very tough and hard wearing, colorfast and stain resistant. The lightweight material — it has the lightest "specific gravity" of any other fabric — dries quickly and wicks sweat and water from the skin. Materials made with Olefin holds its shape extremely well, making it good for packing and active use. It also produces very little static, and weathers well in the sun, as long as special stabilizers are added during manufacturing. Olefin is also highly resistant to deterioration from chemicals and moisture. One of the drawbacks of Olefin is that, because it has a low melting point, it is flammable and will melt if exposed to high heat.

Olefin fibres (polypropylene and polyethylene) are products of the polymerization of propylene and ethylene gases. For the products to be of use as fibres, polymerization must be carried out under controlled conditions with special catalysts that give chains with few branches. Olefin fibres are characterized by their resistance to moisture and chemicals. Of the two, polypropylene is the more favored for general textile applications because of its higher melting point; and the use of polypropylene has progressed rapidly since its introduction. The fibres resist dyeing, so colored olefin fibres are produced by adding dye directly to the polymer prior to or during melt spinning. A range of characteristics can be imparted to olefin fibres with additives, variations in the polymer, and by use of different process conditions.

Olefin Fibre Characteristics

- Able to give good bulk and cover
- Abrasion resistant
- Colorfast
- Quick drying
- Low static
- Resistant to deterioration from chemicals, mildew, perspiration, rot and weather
- Thermally bondable
- Stain and soil resistant
- Strong
- Sunlight resistant
- Dry hand; wicks body moisture from the skin
- Very comfortable
- Very lightweight *(olefin fibres have the lowest specific gravity of all fibres)*

Some Major Olefin Fibre Uses

- **Apparel:** Activewear and sportswear; socks; thermal underwear; lining fabrics
- **Automotive:** Interior fabrics used in or on kick panel, package shelf, seat construction, truck liners, load decks, etc.
- **Home Furnishings:** Indoor and outdoor carpets; carpet backing; upholstery and wall coverings; furniture and bedding construction fabrics
- **Industrial:** Carpets; disposable, durable nonwoven fabrics; ropes; filter fabrics; bagging; geotextiles

General Olefin Fibre Care Tips

- Most stains on fabric can be readily removed by spotting with lukewarm water and detergent.
- Bleaches can be used if needed.
- If fabric is laundered, it should be line dried or tumble dried with gentle heat or no heat. Olefin dries very rapidly.
- Do not iron. *(For specific instructions, refer to garment's sewn-in care label.)*

First U.S. Commercial Olefin Fibre Production: 1958, olefin monofilaments for various specialized uses; 1961, Hercules Incorporated, textile grade multifilament polypropylene

Federal Trade Commission Definition for Olefin Fibre: A manufactured fibre in which the fibre forming substance is any long-chain synthetic polymer composed of at least 85% by weight of ethylene, propylene, or other olefin units, except amorphous *(non-crystalline)* polyolefins.

Olefin fibre is a synthetic fibre made from alkenes. It is used in the manufacture of various textiles as well as clothing, upholstery, wallpaper, ropes, and vehicle interiors. Olefin is also referred to as polypropylene, polyethylene, or polyolefin. Olefin's advantages are its strength, colorfastness and comfort, stain, mildew, abrasion and sunlight resistance, and good bulk and cover.

In organic chemistry, an **alkene**, **olefin**, or **olefine** is an unsaturated chemical compound containing at least one carbon-to-carbon double bond. The simplest acyclic alkenes, with only one double bond and no other functional groups, form a homologous series of hydrocarbons with the general formula C_nH_{2n}.

The simplest alkene is ethylene (C_2H_4), which has the International Union of Pure and Applied Chemistry (IUPAC) name *ethene*. Alkenes are also called *olefins* (an archaic synonym, widely used in the petrochemical industry).

Aromatic compounds are often drawn as cyclic alkenes, but their structure and properties are different and they are not considered to be alkenes.

History

Italy began production of olefin fibres in 1957. The chemist Giulio Natta successfully formulated olefin suitable for more textile applications. U.S. production of olefin fibres began in 1960. Olefin fibres account for 16% of all manufactured fibres.

Major Fibre Properties

Olefin fibres have great bulk and cover while having low specific gravity. This means "warmth without the weight."[1] The fibres have low moisture absorption, but they can wick moisture and dry quickly. Olefin is abrasion, stain, sunlight, and chemical resistant. It does not dye well, but has the advantage of being colorfast. Since Olefin has a low melting point, textiles can be thermally bonded. The fibres have the lowest static of all manufactured fibres and a medium luster. One of the most important properties of olefin is its strength. It keeps its strength in wet or dry conditions and is very resilient. The fibre can be produced for strength of different properties.

Production Method

The Federal Trade Commission's official definition of olefin fibre is "A manufactured fibre in which the fibre forming substance is any long-chain synthetic polymer composed of at least 85% by weight of ethylene, propylene, or other olefin units".

Polymerization of propylene and ethylene gases, controlled with special catalysts, creates olefin fibres. Dye is added directly to the polymer before melt spinning is applied. Additives, polymer variations and different process conditions can create a range of characteristics.

High pressure production, which uses ten tons per square inch, creates a film for molded materials. Low pressure production uses a low temperature with a catalyst and hydrocarbon solvent. This process is less expensive and produces a polyethylene polymer more suitable for textile use.

The polymer is then melted, spun into water, or air cooled. The fibre is drawn out to six times the spun length. Gel spinning is a new method in which a gel form of polyethylene polymers is used.

Olefin fibre is a manufactured fibre in which the fibre-forming substance is any long-chain synthetic polymer composed of at least 85% by weight of ethylene, propylene, or other olefin units. Olefin fibre is a generic description that covers thermoplastic fibres derived from olefins, predominately aliphatic hydrocarbons. Olefins are products of the polymerization of propylene and ethylene gases. Polypropylene (PP) and polyethylene (PE) are the two most common members of the family. Polypropylene is extremely versatile as a fibre-forming material, whereas polyethylene is not as good a fibre-forming high polymer material.

Of the polypropylene used in the U.S., more than one-third goes into fibre and fibre-related products. The declining price and continuing improvements in the quality of polypropylene resins, plus the low specific gravity of the polymer (which provides high covering power), have been important factors in the development of new product end uses.

1. Production Technology For Polyolefins

High Pressure Process For Branched Ldpe

- Autoclave reactor process
- Tubular reactor process

Polymerization Process for Linear Hdpe, Mdpe, Lldpe and Copolymers

- High-pressure autoclave reactor

- Low-pressure liquid slurry processes
- Low and medium pressure solution processes
- Low pressure gas phase processes

Polymerization Process For Pp And Propylene Copolymers

All the above processes have been used in the production of PP. The special process Technologies for PP and propylene copolymers include two kinds of liquid pool slurry Process:

- Low pressure liquid pool slurry phase processes
- Low pressure modular gas phase rectory

Polymerization Process of Other Polyolefins

- Slurry phase heavy-diluents stirred-tank reactor
- Slurry phase light-diluents stirred-tank reactor
- Tubular high-pressure process
- Solution phase medium-pressure adiabatic reactor

2. Processing Methods Of Polyolefins

There are six important processing methods for Polyolefins, these are:

- Injection molding
- Rotational molding
- Blow molding
- Extrusion
- Blown film extrusion
- Cast film extrusion

Processability of Polypropylene (Pp)

A major virtue of polypropylene is its ability to be used in a wide range of fibrous forms. Fibrous forms of polypropylene include staple, bicomponent staple, monofilament, multifilament, slit film yarns, slit-fibrillated film yarns, spunbonds, melt blown nonwovens, synthetic pulps, and extruded nettings. It can be made into ropes and cordage, primary and secondary carpet backing, carpet face yarns, upholstery fabrics, geotextiles, filtration materials, horticulture/ agriculture materials, automotive fabrics, spill-cleanup materials, disposable diapers, hospital/medical care materials, and protective clothing. The melting point of polypropylene (160-170°C) is an advantage in many nonwovens processing steps. PP fibre can be softened sufficiently to bond to one another without destroying fibre properties. Nonwoven fibres made from polypropylene can therefore be fusion-bonded, eliminating the need for chemical binders. The benefits of this technique include both energy saving and environmentally friendliness. Uses of thermally bonded cover stock in baby diapers and similar products will result in a markedly increased use of polypropylene. The fusion characteristics of polypropylene are used not only to bond carded webs but also to improve the dimensional stability of needle-bonded fabrics. A large amount of engineered fabrics for road stabilization, dam and lake reinforcement, soil stabilization and roofing are made from polypropylene fibres. Melt-blown fabrics are widely used in filtration media, battery separators, etc. The relatively high melting point allows PP nonwovens to be used up to the temperature of 120°C before softening occurs. The soft hand and hydrophobic properties make PP nonwovens particularly suitable for hygiene products, baby diapers and adult incontinents. Spunbond and meltblown are the two main processes for polypropylene nonwoven fabrication. Both techniques require PP resins with high melting flow rate and relatively very narrow molecular weight distribution. The fibres produced in spunbonded nonwovens are spun filaments, whose diameters are in the range of 10-35 microns, whereas the fibres of meltblown nonwovens are usually discontinuous and much finer, typically less than 10 microns. This partially explains why meltblown webs are usually weaker than spunbonded webs. Processability of a polymer is highly dependent on its rheological properties, which have close relationship with its molecular weight, molecular weight distribution, and temperature and shear rate. PP resins are generally categorized according to their melt flow rates (MFR), which is the amount of material that passes through a standard die hole for ten minutes. Polymers with higher molecular weight have lower MFR and higher viscosity (under a given temperature). Commercial polypropylene has a wide range of MFR from 0.25 to 1200. MFR is a very important parameter for both melt-blown and spunbond processing. PP melts exhibit non-Newtonian viscosity, normal stress in shear flow, excessive entrance and exit pressure drop, die swell, melt fracture and draw resonance. PP melts are more viscoelastic than PET and nylon melts. The flow pattern and stability of PP melts are highly dependent on the shear rate. Above the critical shear rate, melt fracture may occur.

Processability of polypropylene fibre is also influenced by the die geometry. The L/D ratio has to be optimized to reduce instability and the effects of PP's high viscosity. Both melt fracture and draw resonance represent instabilities in flow. Draw resonance is a periodic variation in diameter of a spinning threadline above a critical draw-down ratio. Slowing down the drawing operation or a suitable cooling procedure may prevent this. In addition, processability of polypropylene fibre can also be affected by other factors such as finishing. Finish oil is a mixture of several chemicals that function as anti-static agent and lubricator to protect the filament. The results of the effects of finishing on this area are not available in this report yet. Since unmodified polypropylene is not dyeable, pigmentation has become the preferred way of coloration in textile and textile-related applications. Generally, the fastness properties of pigmented fibres are superior to those of dyed fibres. Pigmented polypropylene fibres have become established for contract carpets, indoor/outdoor carpeting, synthetic turf, and other applications, in part because of their superior fastness properties. Light stabilizers have helped to open new markets for products intended for use outdoors, and improved heat stabilizers have boosted extrusion efficiency and allowed use of more recycled material.

3. Characteristics of Olefin Fibers

- Good bulk and cover, very lightweight (olefin fibres have the lowest specific gravity of all fibres)
- High strength (wet or dry)
- Resistant to deterioration from chemicals, mildew, insects, perspiration, rot and weather
- Abrasion resistant
- Low moisture absorption
- Stain and soil resistant
- Lowest static component of any man-made fibre
- Sunlight resistant
- Good washability, quick drying, unique wicking
- Resilient, moldable, very comfortable
- Thermally bondable

4. Manufacture of Pp Fiber/Filament

Polypropylene chips can be converted to fibre/filament by traditional melt spinning, though the operating parameters need to be adjusted depending on the final products. Spunbonded and melt blown processes are also very important fibre producing techniques for nonwovens. As an example, the staple fibre production is shown in Fig. 2.

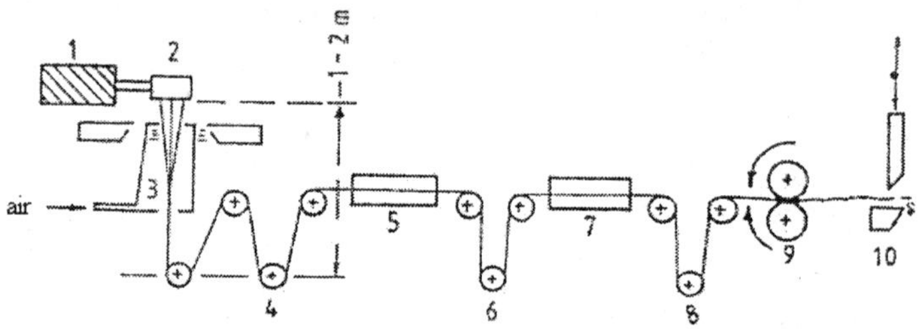

1-extruder, 2-Spinning pack, 3-Ouench duct.4.6.8-Drawing.Tensioning, 5-Hot streching,7-stabilizing,9-Crimper 10-cutter

Fig.2: Schematic diagram of staple fibre production unit of PP.

Identifiers are shown with the figure. Additional comments and description are as follows:

1. Extrusion: L/D=30, compression ratio=1:3.5
2. Metering: one or more spinning gear pumps receives the molten polymer and sends it through the spinning pack to homogenize the product, feed the spinning pack at a constant rate, and prevent fluctuation due to screw extruder.
3. Spinning: the spinning pack consists of three parts-filters, distributor (which distributes the molten polymer over to die surface) and the die. The diameter of the die varies from 0.5 to 1.5mm, depending on the denier required.
4. Quenching: newly extruded filaments are cooled in a good " box" which will distribute 3 m^3/min of cool air without damaging the filaments.
5. Finishing: to improve antistatic and reduce abrasion.
6. Hot Stretching: to enhance the physico-mechanical properties.
7. Crimping: to improve the bulk.
8. Thermosetting: it is a treatment in hot air or steam that removes the internal stresses and relaxes fibres. The resultant fibres are heat-set with increased denier.
9. Cutting: fibres are cut into 20 to 120 mm length depending on whether they are intended for cotton or woolen system.

5. Mechanical Properties

The general physical properties of PP fibres are shown in Table 1. Polypropylene fibres are produced in a variety of types with different tenacities designed to suit varying market requirements. Fibres for general textile uses have tenacities in the range of 4.5-6.0 g/den. High tenacity yarns up to 9.0 g/den are produced for the use in ropes, nets and other similar applications. High performance PP fibres have been made with high strength and high modulus. The techniques include ultra-drawing, solid-state extrusion and crystal surface growth. The filaments with tenacities over 13.0 g/den can be made.

Table 1. : Physical properties of Polypropylene

Moisture Regain	< 0.1%
Refractive Index	1.49
Thermal Conductivity	0.95 Btu-in/ ft ^{2} .hr.°F
Coefficient of linear thermal expansion	4.0x10 $^{-5}$ /°F
Heat of fusion	21 cal/g
Specific heat	0.46 cal/g.c
Density of Melt at 180°c	0.769 g/cc
Heat of Combustion	19,400 Btu/lb
Oxygen index	17.4
Decomposition temperature range	328-410°C
Dielectric constant (0.1 MHz)	2.25
Dissipation factor (0.1 MHz)	< 0.0002
Specific volume resistively	> 10 ^{16} Ω. Cm

The degree of orientation achieved by drawing influences the mechanical properties of PP filaments. The greater the degree of stretch, the higher the tensile strength and the lower the elongation. Commercial PP monofilaments have an elongation-at-break in the range of 12-25%. Multifilaments and staple fibres are in the range of 20-30% and 20-35%, respectively.

6. Thermal Properties

Polypropylene fibres have a softening point in the region of 150°C and a melting point at 160-170°C. At temperatures of -70°C or lower, PP fibres retain their excellent flexibility. At higher temperature (but below 120°C) PP fibres nearly remain their normal mechanical properties. PP fibres have the lowest thermal conductivity of all commercial fibres. In this respect, it is the warmest fibre of all. The thermal conductivity of common textile fibres is shown in Table 2.

Table 2. : Thermal Conductivity of textile fibres

Material	Thermal Conductivity
Air	1.0
PP	6.0
Wool	6.4
Acetate	8.6
Viscose	11
Cotton	17.0

Dyeability

The dyeability of fibres is controlled by both chemical and physical properties. Fibres that have polar functional groups in their molecular repeat units can more easily be dyed. These polar groups may serve as active sites for combination with dye molecules by chemical bonds. Since PP molecular chains have no polar functional groups, the dye molecules cannot be chemically attracted to the fibres. The dye molecules cannot even be strongly adsorbed by the fibre surfaces because of their hydrophobic properties. Therefore, these adsorbed molecules, which interact with the fibre surface by weak van der Waal's force, are easily washed away. In addition, PP fibre has relatively high crystallinity (50-65%), which partially attributes to its very low dyeability.

Fibre modifications have been investigated to make polypropylene dyeable this includes copolymerization to provide dyesites along the macromolecular chain and blend with dyeable polymers. Baumann has reviewed the attempts by several researchers to improve polypropylene dyeability. The method of grafting chloromethylstyrene has been introduced. In addition, the method of blending a polyethyleneimine compound with polypropylene was described briefly. Bromination of polypropylene fibres was reported to give PP fibres an affinity for basic dyes. These methods have, owever, generally led to the significant decrease of the desirable physical properties of polypropylene fibres.

In recent years, surface modification of hydrophobic fibres by plasma treatment has opened up new possibilities in this field. Functional groups can be introduced onto the surface of PP fibres by special gas plasma treatment or plasma polymerization. The polar layer on the fibre surface improves the wettability and the affinity of the fibre surface for water and water-soluble dye molecules. After reaction with the functional groups, which are introduced by plasma treatment, dye molecules are fixed onto the fibre surface. Because plasmas do not interact with the inner part of the fibre, this technique may not greatly influence the mechanical properties of the treated fibre. Great progress in improving dyeability of PP nonwoven webs had been made by the Textiles and Nonwovens development Center (TANDEC), University of Tennessee at Knoxville.

Other Properties of Pp

In general, PP fibre has excellent chemical resistance to acids and alkalis, high abrasion resistance and resistance to insects and pests. PP fibre is also easy to process and inexpensive compared to other synthetic fibres. Its low moisture absorption helps aid the quick transport of moisture.

Disadvantages of Pp Fibres

The main drawbacks of PP fibres are listed below:

- Low melting temperature which prevents it from being ironed like cotton, wool, nylon etc.,
- Hard to be dyed after manufacturing, except after substantial treatment and modification,

- High crystallinity and poor thermal conductivity leads to limited texturizability. [Drawn polypropylene requires a contact time of 2 seconds in the heater compared to PET (POY) which requires only 0.4 seconds]
- Poor UV and thermal stability which requires addition of expensive UV stabilizers and antioxidants to overcome this problem,
- Poor resilience compared to PET and Nylon,
- Creeping due to its low Tg(-15 to -20°C),
- Poor adhesion to glues and latex, and
- Flammable which melts and burns like wax.

Note: PP fibre has many useful textile properties; however, one key disadvantage is that it is flammable. Topical FR treatments have serious deficiencies, such as high add-on and limited durability. Dr. David Buszard (FMC) introduced Reoflam FG-372, a new phosphorus-bromine based melt-processable flame retardant, stable at processing temperatures up to 230°C. It acts as a plasticier during fibre spinning, because of a similar melting temperature. An add-on level of 3% to 8% is required to meet the majority of textile flammability requirements. PP fibres containing Reoflam FG-372 are already being used in automotive carpeting, where good light stability is also of importance.

7. Applications

Because of its superior performance characteristics and comparatively low-cost, PP fibre finds extensive use in the nonwovens industry. PP is a very important fibre in nonwoven processing and dominates in many nonwoven markets. The main application areas include: nonwoven fabrics (refer to Table 3), particularly absorbent product coverstock markets, home furnishings and automotive markets.

Table 3. : Application of Polypropylene Nonwoven fabrics.

Application	Fiber Grade	Industry
Cigarette Filter	Staple fiber 3 denier	Cigarette
Technical filters	Staple fiber 5 dsenier, needle punched nonwoven	Wet filteration, excellent, chemical resistance, used in water, milk, bear, paints, coatings, petrochemicals,Pharmaceuticals, filtration
PP woven socks	PP film fiber, with 10-15% LDPE to reduce fibrillation and cost	Fertilizers, flour, wheat, sugar, cement
Ropes and Twines	PP film and fiber	Agriculture
PP bale warp	Spun Bonded PP	Synthetic fibers
PP tapes	High modulus PP obtained by increasing draw ratio	Construction material like asphalt and concrete
PP construction / industry fabrics	Filling grade and staple fiber	Construction materials like asphalt and concrete
Substrate fabrics	Nonwoven needle punched 3-4 denier staple fibers	Furniture fabrics as backing material for visual furniture fabrics, it serves as reinforcement. Also used for wall covering, luggage, table-clothes, tarpaulins, and automobile
Outdoor Applications	Heavy deniers containing stabilizers, UV absorber, etc	Sports

Non-electric fuses for initiating explosives	PP slit film tapes	Mining industry
Medical/Surgical disposable fabric	PP staple fiber nonwovens, Face masks	Hospital

Below are some major olefin fibre uses:

Carpet and Upholstery

Since polypropylene became a commercially available fibre more than 40 years ago, it has historically been a fibre for carpet and upholstery. About 90% of all carpet backing and more than 25% of all carpet face fibre is polypropylene. In 1997 almost 3.8 billion pounds - approximately 39 percent of all fibres shipped by domestic manufacturers - were delivered to U.S. and Canadian carpet mills. Roughly 1.6 billion pounds of these shipments, 42 percent, were polypropylene filament and staple, representing more than 61 percent of all polypropylene shipments. Table 4 details the growth of polypropylene filament staple distribution since 1993. In 1998, 44% polypropylene filament shipments were used as carpet-face fibres, with slightly more than half this amount, 514 million pounds or 24 percent of all filament shipments, were consumed in backing materials.

Table 4. : Polypropylene Staple Fibre Shipments (millions of pounds)

	1993	1994	1995	1996	1997	1998
Carpet-face yarns	579.6	722.3	820.8	891.4	925.6	935
Carpet-backing yarns	454	485.6	405.7	478.2	488.2	514
Broadwoven	419.9	446.2	448	231.1	235.8	274
Narrow Wovens	19.6	21.3	21.8	15.8	16.5	18

Rope/Cordage/Fishline	113	114.1	121.6	118	111.4	130
Other	45.4	48.7	61.8	197.8	239	240
Total	1,631.5	1,838.2	1,879.7	1,932.3	2,016.5	2,111

Source: Fibre Economics Bureau and Author's Estimates

Polypropylene BCF (bulked continuous filament) yarns now account for almost 1 billion lbs/yr of the face fibre used in carpets and rugs in the U.S. Carpet mills that have back-integrated into fibre production make most of the BCF yarns used in the U.S. carpet industry. In 1998, the relatively low prices for polypropylene resins were a major factor in increased earnings for the major U.S. carpet and rug producers.

Absorbent Products (Diapers)

Absorbent products are very important in the nonwovens business. Of the absorbent product applications, the baby diaper area is the largest volume user, however, applications in adult incontinence currently show the highest growth in recent years.

The major structural components of current diapers are.

- Top sheet (coverstock).
- Acquisition and/or transport or distribution layer.
- Absorbent core.
- Backsheet.

Secondary component materials are:

- Barrier leg cuffs.
- Elastomeric materials.
- Hot melt glues.

Current nonwoven materials used in top sheet applications include spunbonded polypropylene (usually produced on a multibeam system), SMS (spunbond/melt blown/spunbond) polypropylene composites and carded polypropylene thermal bonds. In some applications, thermally bonded bicomponent structures are being used.

Polypropylene spunbonded webs, treated to render them hydrophilic (or partially treated to make them hydrophilic in zones) have proved popular materials for coverstock applications, but they have not always been rated as soft as other materials, i.e., staple fibre-based thermally bonded polypropylene.

Automotive Products

The nonwovens business in South America has continued to expand in more and more automotive product segments despite an automotive industry crisis. The past year saw only 1.6 million units produced, a 24% drop from 1997. Sales during the year fell 21% and exports dropped 7.7%, from 416,000 to 384,000 units. Imports were another story, with new car imports growing 13.8%, from 303,000 to 345,000 in numbers of units. The negative trends in 1998 can largely be attributed to a combination of economic problems, increased unemployment levels, higher interest rates and tributary charges.

Principal polypropylene applications for nonwovens in automobiles are interior fabrics used in or on kick panel, package shelf, seat construction, truck liners, load decks, cabin air filters etc

Physical and chemical structure

Physical

Olefin fibres can be multi- or monofilament and staple, tow or film yarns. The fibres are colorless and round in cross section. This cross section can be modified for different end uses. The physical characteristics are a waxy feel and colorless.

Chemical

There are two types of polymers that can be used in

olefin fibres. The first, polyethylene, is a simple linear structure with repeating units. These fibres are used mainly for ropes, twines and utility fabrics.

The second type, polypropylene, is a three dimensional structure with a backbone of carbon atoms. Methyl groups protrude from this backbone. Stereo selective polymerization orders these methyl groups to the same spatial placement. This creates a crystalline polypropylene polymer. The fibres made with these polymers can be used in apparel, furnishing and industrial products.

Manufacturers

The first commercial producer of an olefin fibre in the United States was Hercules, Inc. (Fibre Visions). Other U.S. olefin fibre producers include Asota; American Fibres and Yarns Co; American Synthetic Fibre, LLC; Color-Fi; Fibre Visions; Foss Manufacturing Co., LLC; Drake Extrusion; Filament Fibre Technology, Inc.; TenCate Geosynthetics; Universal Fibre Systems LLC.

Trademarks according to fabric use

Producer – Allied-Signal

- A.C.E. – Tire cord, furniture webbing

Producer – DuPont

- CoolMax – Warm-weather and action wear
- Hollofil, Quallofil – Fibrefill and insulating fibres
- Sontara – Spunlaced nonwoven fabrics
- Thermostat – Cold-weather wear
- Thermoloft - Fibrefill and insulating fibres
- Tyvek - Used for house wraps to postal envelopes to clothing

Producer – Trevira

- ESP – Apparel and furnishings
- Celwet – Nonwovens
- Comfort Fibre – Staple fibre for apparel uses
- Loftguard – Staple fibre for industrial uses

- Polar Guard
- Lambda – Filament yarns with spun-yarn characteristics
- Serene
- Superba
- Trevira HT – Marine and military uses; ropes, cordages
- Trevira ProEarth – Recycled-content geotextiles
- Trevira XPS - Carpeting
- BTU – Cold-weather apparel

Producer - 3M

- Thinsulate - Cold-weather action wear

Uses

Apparel

Sports & active wear, socks, thermal underwear; lining fabrics.

Home Furnishing

Olefin can be used by itself or in blends for indoor and outdoor carpets and carpet tiles, carpet backing. The fibre can also be used in upholstery, draperies, wall coverings, slipcovers, and floor coverings.

Automotive

Olefin can be used for interior fabrics, sun visors, arm rests, door and side panels, trunks, parcel shelfs, and resin replacement as binder fibres.

Industrial

In an industrial setting, olefin creates carpets; ropes, geotextiles that are in contact with the soil, filter fabrics, bagging, concrete reinforcement, and heat-sealable paper (e.g. tea- and coffee-bags).

Care procedures

Dry cleaning Olefin is not recommended, because many dry-cleaning solvents can swell the fibres. As Olefin dries quickly, line drying and low tumble drying with little or no heat is the recommended method of drying. Since Olefin is not absorbent, waterborne stains do not present a problem.

However, oily stains are difficult to remove. Most such stains can be removed with lukewarm water and detergent, but bleach can also be used. Olefin fibre has a low melting point (around 225 to 335 °F, depending on the polymer's grade) so items should be ironed at a very low temperature, if at all. Items such as outdoor carpets and other fabrics can be hosed off. Olefin is easy to recycle.

8

Polyester

Polyester is one of the major synthetic fibres. Its name must be familiar to most shoppers; 100% polyester is seen often on the labels of ladies and men's outerwear, while the blend polyester/cotton is very common in shirts and bed sheets. Another very common blend, polyester/wool is popular in dress fabrics and men's suits. There are a number of reasons for this wide use of polyester. Most important are its properties of easy care and toughness, while its relatively low price obviously has a marked influence on its universal acceptance.

The raw material for the production of polyester is oil. Polyester is similar to nylon in many respects in that fibres formed from both of them are strong with high abrasion resistance and low moisture absorption. They are also resistant to rot and chemicals and can be set into shapes by the application of heat. Polyester, however, even with the possible disadvantage of its later commercial development, has become probably the most important synthetic fibre. This has happened because of relatively slightly but important differences in its properties compared with those of other synthetics, and production costs lower than those of nylon.

One of the most important properties of the fibre is its resistance to stretching. The handle or feel of the fabric is more influenced by the modulus of the fibres i.e. their resistance to small extensions and bending. The modulus of polyester is about double that of nylon, which tends to give fabrics containing polyester a crisper handle and good dimensional stability. The ability of polyester fabrics, after heat setting, to retain a flat shape by shedding accidental creases but also to retain pleats and fashion creases set into the fabric is a major advantage.

Because of its high melting point polyester has good thermal stability. It softens at about 200ÚC and will not easily burn. Unless supported in a fabric by another fibre it will melt away from heat rather than burst into flame. At a temperature of 150ÚC it retains about 50% of its tenacity and has good long term resistance to heat in the absence of strong chemicals.

Polyester's easy care properties are helped by a very low value of moisture absorption. Under normal atmospheric conditions it absorbs only about 0.4% water compared with 4% for nylon and 7% for cotton. Its strength is little affected when wet and it also has the added advantage of drying very quickly.

World production figures illustrate the dominance of polyester as the current main synthetic fibre. In 1970 nylon had the highest production value of all synthetics, at a level of 40%, with polyester second at 34%. Since 1975, however, when nylon had fallen to 33% and polyester had passed it to a level of 45%, polyester continued to grow on a percentage basis so that in 1998 it had reached 58% of world synthetic fibre production, mainly at the expense of nylon, which had fallen to less than 15% (acrylics account for most of the remaining production). These changes occurred at a time when production shifted from the USA and Western Europe to developing countries, who now produce the lion's share of polyester fibre.

The environmental impact of the production of polyester is relatively small, or at least looks that way, since the polymer once formed is just heated until molten before being formed into fibres with no effluent being produced. However, other factors such as high energy consumption, and process emissions generated in both the production of the constituent chemicals and polymer have also to be considered. The oil required for all of this process being a non-recoverable resource. It can therefore be argued that the production of polyester, and indeed any other synthetic fibre where the polymer is generated from petroleum, is not sustainable. In terms of the world consumption of chemical feedstock then the production of man-made fibres, of which polyester is the greatest, accounts for about 5% of the total.

When dyeing polyester, dye carriers are required to help achieve complete dye penetration. Many of these chemicals tend to be toxic in nature and some are carcinogenic. Those with a higher risk have already been effectively banned from use.

The setting of synthetics such as polyester, produces a characteristic blue stenter haze from the oils being driven off the fabric. The emissions from this process can be quite odorous and so textile finishers are finding that local authorities are insisting that abatement systems are installed to remove the fume.

Polyesters are the polymers, in the form of fibres, that were used back in the seventies to make all that wonderful disco clothing, the kind you see being modeled on the right. But since then, the nations of the world have striven to develop more tasteful uses for polyesters, like those nifty shatterproof plastic bottles that hold your favorite refreshing beverages, like the blue bottle in the picture below. So you see, polyesters can be both plastics and fibres. Another place you find polyester is in balloons. Not the cheap ones that you use for water balloons, those are made of natural rubber. I'm talking about the fancy ones you get when you're in the hospital. These are made of a polyester film made by DuPont called Mylar. The balloons are made of a sandwich, composed of Mylar and aluminum foil. Materials like this, made of two kinds of material, are called composites.

Polyester is a category of polymers which contain the ester functional group in their main chain. Although there are many polyesters, the term "polyester" as a specific material most commonly refers to polyethylene terephthalate (PET). Polyesters include naturally-occurring chemicals, such as in the cutin of plant cuticles, as well as synthetics such as polycarbonate and polybutyrate. Natural polyesters and a few synthetic ones are biodegradable, but most synthetic polyesters are not.

Polyesters may be produced in numerous forms such as sheets and three-dimensional shapes. Polyesters as thermoplastics may change shape after the application of heat. While combustible at high temperatures, polyesters tend to shrink away from flames and self-extinguish upon ignition. Polyester fibres have high tenacity and E-modulus as well as low water absorption and minimal shrinkage in comparison with other industrial fibres.

Woven polyester fabrics are used in apparel and home furnishings such as bed sheets, beds, curtains and draperies. Similarly, industrial polyesters are used in tyre reinforcements, ropes, fabrics for conveyor belts, safety belts, coated fabrics and plastic reinforcements with high energy absorption. Polyester fibre fills are also used to stuff pillows, comforters and cushion padding.

Polyester fabrics are claimed to have a "less natural" feel when compared to similarly-woven fabrics made from natural fibres (i.e. cotton in textile uses). However, polyester fabrics may exhibit other advantages over natural fabrics, such as improved wrinkle resistance. As a result, polyester fibres are sometimes spun together with natural fibres to produce a cloth with blended properties.

Polyesters are also used to make bottles, films, tarpaulin, canoes, liquid crystal displays, holograms, filters, dielectric film for capacitors, film insulation for wire and insulating tapes.

Liquid crystalline polyesters are among the first industrially-used liquid crystalline polymers. They are used for their mechanical properties and heat-resistance. These traits also important in their application as an abradable seal in jet engines.

Thermosetting polyesters are used as casting materials, and chemosetting polyester resins are used as fibreglass laminating resins and non-metallic auto-body fillers. Fibreglass-reinforced unsaturated polyesters find wide application in bodies of yachts and as body parts of cars.

Polyesters are also widely used as a finish on high-quality wood products such as guitars, pianos and vehicle / yacht interiors. Burns Guitars, Rolls Royce and Sunseeker are a few companies that use polyesters to finish their products. Thixotropic properties of spray-applicable polyesters make them ideal for use on open-grain timbers, as they can quickly fill wood grain, with a high-build film thickness per coat. Cured polyesters can be sanded and polished to a high-gloss, durable finish.

A special family of polyesters are polycarbonates. Polyesters have hydrocarbon backbones which contain ester linkages, hence the name.

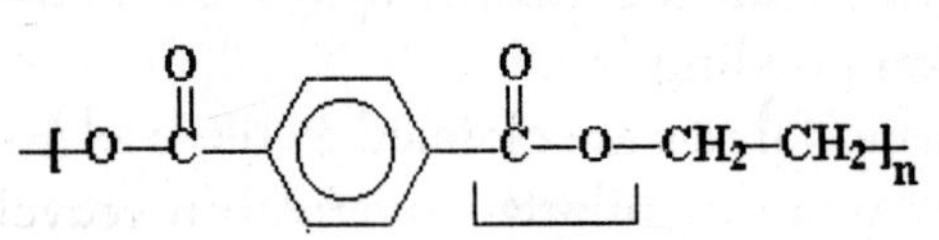

The structure in the picture is called poly (ethylene terephthalate), or PET for short, because it is made up of ethylene groups and terephthalate groups (duh!). I realize that *terephthalate* is not the kind of word most English-speaking mouths are used to saying, but with practice you should be able to say it with only a slight feeling of awkwardness when it rolls off your tongue.

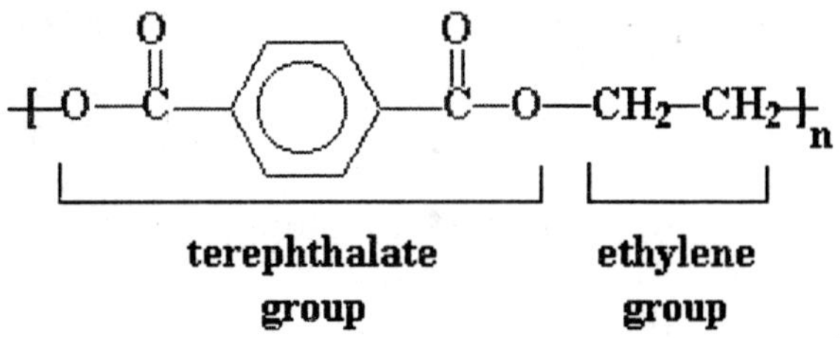

The ester groups in the polyester chain are polar, with the carbonyl oxygen atom having a somewhat negative charge and the carbonyl carbon atom having a somewhat positive charge. The positive and negative charges of different ester groups are attracted to each other. This allows the ester groups of nearby chains to line up with each other in crystal form, which is why they can form strong fibres.

Polyester is a manufactured product made from synthesized polymers. It tends to be very resilient, quick drying, resistant to biological damage such as mold and mildew, easy to wash, and able to hold forms well. While Polyester is often maligned as a textile, it has many useful applications. Polyester is, however, highly flammable, so care should be taken when wearing it. Most synthetic fabrics are subject to flammability by nature, because they are made from polymers.

Polyester is not just a textile. It is made from polyethylene terephthalate (PET), the same material used to make plastic drink bottles. Many drink bottles are recycled by being reheated and turned into Polyester fibres, which in addition to being an efficient use, also helps keep polymers out of landfills. Polyester is a plastic, invented in Britain in the early 1940s by two gentlemen working with synthetic polymers. In the 1950s, the new textile took off, becoming popular for its easy care, drape, and versatility.

In the United States, Dupont is one of the major manufacturers of PET and has been making Dacron Polyester since 1950. Dupont has researched the properties of Polyester extensively and devised a number of uses well beyond textiles, such as mylar and other Polyester films. Dupont has also been driven to find ways to recycle Polyester products to help keep costs down, as the base chemical substances that combine to form Polyester are petroleum products.

To make Polyester, ethylene glycol and dimethyl terephthalate are mixed together. The chemical reaction results in bisterephthalate. This substance is heated to 270° Fahrenheit (132° Celsius), where it reacts again to form polyethylene terephthalate. Like many chemical reactions that result in polymers, the Polyester making process results in unhealthy off gassing, and protection should be worn while making PET. After synthesizing the polymers, the manufacturer decides what to do with them.

PET can be formed into plastics that can later be recycled. It is a highly malleable material and appears in all sorts of applications, from drink bottles to floppy disk liners. PET can also be used to make Polyester fibre, which is used in auto upholstery, quilt batting, and clothing of all sorts.

To make Polyester fibre, an extruder is used to produce very fine threads of PET. Polyester clothing tends to be slippery and silky in feel, although it can cause skin irritation for some wearers. The Polyester fibres used to make clothing can be knitted or woven, although most are knit, to maximize the flexibility of Polyester. Some Polyester is blended with other fabrics to provide more loft or stretch, or to minimize skin irritation.

History of Polyester

British chemists, John Rex Whinfield and James Tennant

Dickson, employees of the Calico Printer's Association of Manchester, patented "polyethylene terephthalate" (also called PET or PETE) in 1941, after advancing the early research of Wallace Carothers. The saw that Carothers's research had not investigated the polyester formed from ethylene glycol and terephthalic acid. Polyethylene terephthalate is the basis of synthetic fibres such as polyester, dacron, and terylene. Whinfield and Dickson along with inventors W.K. Birtwhistle and C.G. Ritchiethey also created the first polyester fibre called Terylene in 1941 (first manufactured by Imperial Chemical Industries or ICI). The second polyester fibre was Dupont's Dacron.

According to Dupont, "In the late 1920s, DuPont was in direct competition with Britain's recently formed Imperial Chemical Industries. DuPont and ICI agreed in October 1929 to share information about patents and research developments. In 1952, the companies' alliance was dissolved... The polymer that became polyester has roots in the 1929 writings of Wallace Carothers. However, DuPont chose to concentrate on the more promising nylon research. When DuPont resumed its polyester research, ICI had patented Terylene polyester, to which DuPont purchased the U.S. rights in 1945 for further development. In 1950, a pilot plant at the Seaford, Delaware, facility produced Dacron [polyester] fibre with modified nylon technology."

Dupont's polyester research lead to a whole range of trademarked products, one example is Mylar (1952), an extraordinarily strong polyester (PET) film that grew out of the development of Dacron in the early 1950s.

Polyesters are made from chemical substances found mainly in petroleum and are manufactured in fibres, films, and plastics.

According to Dupont Teijin Films, "Plain polyethylene terephthalate (PET) or polyester is most commonly associated with a material from which cloth and high-performance clothing are produced (e.g., DuPont Dacron® polyester fibre). Increasingly over the last ten years PET has gained acceptance as a material of choice for beverage bottles... PETG, also known as glycolised polyester, is used in the production of cards... Polyester film (PETF) is a semi-crystalline film used in many applications such as videotape, high quality packaging, professional photographic printing, X-ray film, floppy disks, etc. "

DuPont Teijin Films (founded January 1, 2000) is a leading supplier of PET and PEN polyester films whose brand names include: Mylar ®, Melinex ®, and Teijin ® Tetoron ® PET polyester film, Teonex ® PEN polyester film, and Cronar ® polyester photographic base film.

Naming an invention actually involves developing at least two names. One name is the generic name. The other name is the brand name or trademark. For example, Mylar ® and Teijin ® are brand names; polyester film or polyethylene terephthalate are the generic or product names.

Basic Principles of Polyester Fibre Production — The most common polyester for fibre purposes is poly (ethylene terephthalate), or simply PET. This is also the polymer used for many soft drink bottles and it is becoming increasingly common to recycle them after use by remelting the PET and extruding it as fibre. This saves valuable petroleum raw materials, reduces energy consumption, and eliminates solid waste sent to landfills.

PET is made by reacting ethylene glycol with either terephthalic acid or its methyl ester in the presence of an antimony catalyst. The reaction is carried out at high temperature and vacuum to achieve the high molecular weights need to form useful fibres. PET is melt spun.

Generic name for a manufactured fibre in which the fibre-forming substance is any long chain synthetic polymer composed of at least 85 % by weight of an ester of dihydric alcohol and terephthalic acid which is derived from polyethylene terephthalate. Polyester filaments are produced by forcing the molten polymer at a temperature of about 290°C through spinneret holes followed by air cooling, then combining the single filaments into yarns, and drawing.

Polyester Fibre Characteristics

- Strong
- Resistant to stretching and shrinking
- Resistant to most chemicals
- Quick drying
- Crisp and resilient when wet or dry
- Wrinkle resistant
- Mildew resistant
- Abrasion resistant
- Retains heat-set pleats and crease
- Easily washed

Some Major Polyester Fibre Uses

- **Apparel:** Every form of clothing
- **Home Furnishings:** Carpets, curtains, draperies, sheets and pillow cases, wall coverings, and upholstery
- **Other Uses:** hoses, power belting, ropes and nets, thread, tire cord, auto upholstery, sails, floppy disk liners, and fibrefill for various products including pillows and furniture

General Polyester Fibre Care Tips

- Most items made from polyester can be machine washed and dried. Use warm water and add a fabric softener to the final rinse cycle. Machine dry at a low temperature and remove articles as soon as the tumbling cycle is completed.
- If ironing is desired, use a moderately warm iron.
- Most items made from polyester can be dry-cleaned. *(For specific instructions, refer to garment's sewn-in care label.)*

Federal Trade Commission Definition for Polyester Fibre: A manufactured fibre in which the fibre forming substance is any long-chain synthetic polymer composed of at least 85% by weight of an ester of a substituted aromatic carboxylic acid, including but not restricted to substituted terephthalic units, $p(\text{-R-O-CO-}\ C_6H_4\text{-CO-O-})_x$ and parasubstituted hydroxy-benzoate units, $p(\text{-R-O-CO-}C_6H_4\text{-O-})_x$.

First U.S. Commercial Polyester Fibre Production: 1953, DuPont Company

The Polyester Industry

Basics : Polyester is a synthetic polymer made of purified terephthalic acid (PTA) or its dimethyl ester dimethyl terephthalate (DMT) and monoethylene glycol (MEG). With 18% market share of all plastic materials produced, it ranges third after polyethylene (33.5%) and polypropylene (19,5%).

The main raw materials are described as follows:

- **Purified Terephthalic Acid – PTA – CAS-No.: 100-21-0**
Synonym: 1,4 Dibenzenedicarboxylic acid,
Sum formula; $C_6H_4(COOH)_2$, mol weight: 166,13
- **Dimethylterephthalate – DMT- CAS-No: 120-61-6**
Synonym: 1,4 Dibenzenedicarboxylic acid dimethyl ester
Sum formula $C_6H_4(COOCH_3)_2$, mol weight: 194,19
- **Mono Ethylene Glycol – MEG – CAS No.: 107-21-1**
Synonym: 1,2 Ethanediol
Sum formula: $C_2H_6O_2$, mol weight: 62,07

More information about polyester raw materials can be found for PTA, DMT and MEG, at the webpage INCHEM "Chemical Safety Information from Intergovernmental Organizations".

To make a polymer of high molecular weight a catalyst is needed. The most common catalyst is antimony trioxide (or antimony tri acetate):

Polyester is described as follows:

Polyethylene Terephthalate CAS-No.: 25038-59-9 Synonym / abbreviations: polyester, PET, PES Sum Formula: H-[C10H8O4]-n=60-120 OH, mol unit weight: 192,17

There are several reasons for the importance of PTA:

- The relatively easy accessible raw materials PTA or DMT and MEG
- The very well understood and described simple chemical process of polyester synthesis
- The low toxicity level of all raw materials and side products during production and processing

- The possibility to produce PET in a closed loop at low emissions to the environment
- The outstanding mechanical and chemical properties of polyester
- The recycle ability
- The wide variety of intermediate and final products made of polyester

In table 1 the estimated world`84

polyester production for textile polyester, bottle polyester resin, film polyester mainly for packaging and specialty polyesters for engineering plastics, which are the main fields of application, can be seen. According to this table, the world's total polyester production might exceed 50 million tons per annum before the year 2010.

Table 1. : World polyester production

Product Type	Market size per year 2002 [Mio t/a]	2008 [Mio t/a]
Textile-PET	20	39
Resin, Bottle/A-PET	9	16
Film-PET	1.2	1.5
Special Polyester	1	2.5
TOTAL	31.2	49

Raw Material Producer

The raw materials PTA, DMT and MEG are mainly produced by large chemical companies which are sometimes integrated down to the crude oil refinery where p-xylene is the base material to produce PTA and liquefied petroleum gas (LPG) is the base material to produce MEG.

Large PTA producers are for instance BP, Reliance, Sinopec, SK-Chemicals, Mitsui and Eastman Chemicals. MEG production is in the hand of about 10 global players which are headed by MEGlobal a JV of DOW and PIC Kuweit followed by Sabic.

Among the world's largest polyester producers are the following companies:

Artenius, Advansa, DAK, DuPont, Eastman/Voridian, Hyosung, Huvis, Indorama, Invista, Jiangsu Hengli Chemical Fibre, Jiangsu Sanfangxian Industry, M&G Group, Mitsui, Mitsubishi, NanYa Plastics, Reichhold, Reliance, Rongsheng, Sabic, Teijin, Toray, Trevira, Tuntex, Wellman, Yizheng Sinopec, Zhejiang Hengi Polymerization.

With more than 500 plants in China, about half of the world production originates in that country. More information about polyester in China can be found under the web site of China Chemical Fibre Economic Information Network.

Polyester Processing

After the first stage of polymer production in the melt phase, the product stream divides into two different application areas which are mainly textile applications and packaging applications. In figure 2 the main applications of textile and packaging polyester are listed.

Textile and packaging polyester application list
Polyester-based Polymer (Melt Or Pellet)

Textile	Packaging
Staple fibre (PSF)	Bottles for CSD, Water, Beer, Juice, Detergents
Filaments POY, DTY, FDY	A-PET Film
Technical yarn and tire cord	Thermoforming
Non-woven and spunbond	BO-PET Biaxial oriented Film
Mono-filament	Strapping

Abbreviations: PSF = Polyester Staple Fibre; POY = Partially Oriented Yarn; DTY = Draw Textured Yarn; FDY = Fully Drawn Yarn; CSD = Carbonated Soft Drink; A-PET =

Amorphous Polyester Film; BO-PET = Biaxial Oriented Polyester Film;

A comparable small market segment (<< 1 Million t/a) of polyester is used to produce engineering plastics and masturbatch.

In order to produce the polyester melt with a high efficiency, high-output processing steps like staple fibre (50–300 t/d per spinning line) or POY /FDY (up to 600 t/d split into about 10 spinning machines) are meanwhile more and more horizontal, integrated, direct processes. This means the polymer melt is directly converted into the textile fibres or filaments without the common step of pelletizing. We are talking about full horizontal integration when polyester is produced at one site starting from crude oil or distillation products in the chain oil -> benzene -> PX -> PTA -> PET melt -> fibre / filament or bottle-grade resin. Such integrated processes are meanwhile established in more or less interrupted processes at one production site. Eastman Chemicals introduced at first the idea to close the chain from PX to PET resin with their so-called INTEGREX process. The capacity of such horizontal, integrated productions sites is >1000 t/d and can easily reach 2500 t/d.

Besides the above mentioned large processing units to produce staple fibre or yarns, there are ten thousands of small and very small processing plants, so that one can estimate that polyester is processed and recycled in more than 10 000 plants around the globe. This is without counting all the companies involved in the supply industry, beginning with engineering and processing machines and ending with special additives, stabilizers and colors. This is a gigantic industry complex and it is still growing by 4–8% per annum, depending on the world region. Useful information about the polyester industry can be found under where a "Who is Producing What in the Polyester Industry" is gradually being developed.

Synthesis

Synthesis of polyesters is generally achieved by a polycondensation reaction. See "condensation reactions in polymer chemistry". The General equation for the reaction of a diol with a diacid is : (n+1) $R(OH)_2$ + n $R'(COOH)_2$ —> $HO[ROOCR'COO]_nROH$ + 2n H_2O

Azeotrope Esterification

In this classical method, an alcohol and a carboxylic acid react to form a carboxylic ester. To assemble a polymer, the water formed by the reaction must be continually removed by azeotrope distillation.

Acylation (HCl method)

The acid begins as an acid chloride, and thus the polycondensation proceeds with emission of hydrochloric acid (HCl) instead of water. This method can be carried out in solution or as an enamel.

Ring-opening Polymerization

Aliphatic polyesters can be assembled from lactones under very mild conditions, catalyzed anionically, cationically or metallorganically.

Thermosetting resins are generally copolymers of unsaturated polyesters with styrene. Polyester saturation is governed through the use of maleic acid or fumaric acid. In vinyl esters, saturation (or lack thereof) is found in the alcohol group of the polyester. The double bond of unsaturated polyester reacts with styrene resulting in a 3-D cross-linked structure. This structure acts as a thermoset. The cross-linking is initiated through an exothermic reaction involving an organic peroxide, such as methyl ethyl ketone peroxide or benzoyl peroxide.

Background

Polyethylene terephthalate polyester (PETP) is the most common thermoplastic polyester and is often called just "polyester".

This often causes confusion - not only is the chemically similar polybutylene terephthalate (PBT) also a (thermoplastic) polyester, the most common resin system used in GRP is also a polyester system - and also often called just "polyester". (In this latter case, however, the polyesters are chemically unsaturated and are "free-radical polymerised" into a thermoset).

General Properties

Polyethylene terephthalate (PET) is a hard, stiff, strong, dimensionally stable material that absorbs very little water. It has good gas barrier properties and good chemical resistance except to alkalis (which hydrolyse it). Its crystallinity varies from amorphous to fairly high crystalline. It can be highly transparent and colourless but thicker sections are usually opaque and off-white.

Polyethylene Terephthalate Films

It is widely known in the form of biaxially oriented and thermally stabilised films usually referred to by their main brand names Mylar, Melinex or Hostaphan. Strictly speaking, these names should be used only for this type of film whose properties are different from, and in several respects superior to, those of "ordinary" polyethylene terephthalate (PET) film.

Applications

These "Mylar®-type" films are used for capacitators, graphics, film base and recording tapes etc. PET is also used for fibres for a very wide range of textile and industrial uses (Dacron®, Trevira®, Terylene®). Other applications include bottles and electrical components.

Key Properties

Chemical Resistance	
Acids – concentrated	Good
Acids – dilute	Good
Alcohols	Good
Alkalis	Poor

Aromatic hydrocarbons	Fair
Greases and Oils	Good
Halogens	Good
Ketones	Good
Electrical Properties	
Dielectric constant @1MHz	3.0
Dielectric strength ($kV.mm^{-1}$)	17
Dissipation factor @ 1kHz	0.002
Surface resistivity (Ohm/sq)	10^{13}
Volume resistivity (Ohm.cm)	$>10^{14}$

Mechanical Properties	
Coefficient of friction	0.2-0.4
Hardness–Rockwell	M94-101
Izod impact strength ($J.m^{-1}$)	13-35
Poisson's ratio	0.37-0.44 (oriented)
Tensile modulus (GPa)	2-4
Tensile strength (MPa)	80, for biax film 190-260

Physical Properties	
Density ($g.cm^{-3}$)	1.3-1.4
Flammability	Self Extinguishing
Limiting oxygen index (%)	21
Refractive index	1.58-1.64
Resistance to Ultra-violet	Good
Water absorption - equilibrium (%)	<0.7
Water absorption-over 24 hours (%)	0.1

Thermal Properties	
Coefficient of thermal expansion ($x10^{-6}$ K^{-1})	20-80
Heat-deflection temperature -0.45MPa (°C)	115
Heat-deflection temperature -1.8MPa (°C)	80
Lower working temperature (°C)	-40 to -60
Specific heat ($J.K^{-1}.kg^{-1}$)	1200-1350
Thermal conductivity ($W.m^{-1}.K^{-1}$)	0.15-0.4 @ 23
Upper working temperature (°C)	115-170

Properties Polyethylene Terephthalate Film

Property		Value
Dielectric Strength @25μm thick	$KV.mm^{-1}$	300
Dissipation Factor @1MHz		0.016
Elongation at Break	%	60-165

Initial Tear Strength	$g.\mu m^{-1}$	18-54
Permeability to Carbon Dioxide @25°C	$x10^{-13}$ $cm^3.cm.cm^{-2}.s^{-1}.Pa^{-1}$	0.07-0.11
Permeability to Hydrogen @25°C	$x10^{-13}$ $cm^3.cm.cm^{-2}.s^{-1}.Pa^{-1}$	0.45
Permeability to Nitrogen @25°C	$x10^{-13}$ $cm^3.cm.cm^{-2}.s^{-1}.Pa^{-1}$	0.0034-0.0038
Permeability to Oxygen @25°C	$X10^{-13}cm^3.cm.cm^{-2}.s^{-1}.Pa^{-1}$	0.015-0.04
Permeability to Water @25°C	$X10^{-13}$ $cm^3.cm.cm^{-2}.s^{-1}.Pa^{-1}$	100-115
Specific Heat	$KJ.kg^{-1}.K^{-1}$	1.3
Thermal Conductivity @23°C	$W.m^{-1}.K^{-1}$	0.13-0.15

Properties Polyethylene Terephthalate Fibre

Property		Value	
Material		Medium	High tenacity
Specific Modulus	cN/tex		700-800
Specific Tenacity	cN/tex	36	70-80
Density	$g.cm^{-3}$	1.39	1.39
Extension to break	%	36	13-16
Modulus	GPa		9-11
Shrinkage @100°C	%	4	1.5-6
Tenacity	GPa	0.5	0.9-1.1

Production

Polyesters are made from chemical substances found mainly in petroluem. Polyesters are manufactured in three basic forms - *fibres, films* and *plastics.*

Polyester fibres are used to make fabrics. Poly (ethylene terephthalate, or simply PET) is the most common polyester used for fibre purposes. This is the polymer used for making soft drink bottles. Recycling PET by re-melting it and extruding it as fibre saves much raw materials as well as energy.

PET is made by ethylene glycol with either terephthalic acid or its methyl ester in the presence of an antimony catalyst. In order to achieve high molecular weights needed to form useful fibres, the reaction has to be carried out at high temperature and in a vacuum.

Care of your polyester garment:

- Machine-wash your fabric in cold water.
- Wash with light colours.
- Use a gentle cycle while washing.
- Tumble dry with the setting low.
- Do not bleach.
- If needed, iron with a cool iron.

Properties of the polyester:

- It is resists wrinkling.
- It is easy to launder.
- It dries quickly.
- It is resistant to stretching and shrinking.

Uses

Polyester is used to make most forms of clothing like shirts, running shorts, track pants, windbreakers, and lingerie. It can also be made into curtains and draperies.

Recycled Polyester

Recycled Polyester is a polyester that has been manufactured by using previously used polyester items. In the clothing world, recycled polyester clothes can be created from used clothes.

Polyester fibres made from recycled plastic containers (such as Fortrel Ecospun) are often referred to as recycled polyester too. However, in this article, I'm referring to polyester clothes made from other polyester clothes.

Recycled polyester is not always eco-friendly. The "eco-friendliness" of recycled polyester depends on the recycling process, and the original polyester itself. Polyester is still a synthetic fabric, so it's not going to be as organic as plant fibres such as hemp, bamboo, ramie etc.

Traditional polyester has potential health hazards and environmental issues that make it a lot less eco-friendly than it could be. A relatively new brand of polyester called Eco-Intelligent Polyester (launched in 2001) aims to overcome the health hazards and environmental issues inherent in traditional polyester.

Traditional synthetic polyester has potential health hazards. This is because antimony is used during the manufacturing stage. Antimony is toxic to the heart, lungs, liver and skin, and manufacture of polyester produces a by-product called antimony trioxide. Over the long term, inhalation of antimony trioxide can cause chronic bronchitis and even emphysema. What's more, the antimony trioxide mixes with the wastewater and you end up with toxic wastewater.

Recycling this class of polyester only exasperates things. The recycling process is a high-temperature process and you end up with antimony trioxide again. Plus, the recycling process actually results in a lower quality polyester. This is why the term "downcycling" is more appropriate than "recycling". Over time, continued recycling leads to a polyester with less and less value - until it ends up in landfill.

Eco-Intelligent Polyester

With the inherent environmental and health issues related to traditional polyester recycling, the whole process needed to be redesigned. With the traditional method, the problem started with the original polyester. This is the first thing that needed to be changed. You may have heard the term "garbage in, garbage out". Well, that term could also apply with recycling.

For the best recycling result, all ingredients of the original polyester should be healthy and environmentally friendly. Every ingredient needs to have the ability to be recycled into something of at least the same quality. That way, a new (and environmentally friendly) recycling process can be created that maintains the quality of the original product, and is friendly to the environment.

Virgin pet resin competes directly with recycled PET for many applications. Unfortunately, virgin PET can undercut reclaimed resin in price (under certain circumstances), has higher quality over its recycled counterpart, and is not subject to FDA restrictions on its usage. These disadvantages create quite an uphill battle for recycled PET use. However, many companies have found new ways to let competition further benefit them.

Alternatives for PET Recyclers

Today's PET recyclers work to add value to their product through vertical integration into fibres, thermoformable sheet, and compounding into alloyed products, fig 1. This not only creates a stronger combined product, but also significantly increases the value of the resin, and avoids direct competition with virgin PET. By integrating the recovery of PET into downstream manufacturing processes, market stability, product quality, and profits have increased greatly. With the addition of these processes to the PET recycling industry, competition with the powerful virgin resin producers will decrease and prosperity in the industry will increase.

The Recycled PET Market

In the sports world, competing against those who have better skills than you is said to greatly help your play. As you continue to be knocked around, you gain better skill and knowledge of the game. In the PET recycling industry, however, this competition has not been as beneficial to all the players involved. Because of large oversupply of virgin PET, today's PET recyclers have lost three quarters of their industry and have been forced to limit collection of PET bottles.

Plastic Container Recovery

According to the National Association for Plastic Container Recovery (Charlotte, North Carolina), the US recovered 572 million pounds of PET in 1996. The PET recycling industry was hit hard in 1996 by an oversupply of PET virgin material that pushed prices for all PET material down, including recycled PET. More than three -quarters of the PET recyclers processing in 1995 were forced to shutdown in 1996 because of these difficult market conditions.

Price Fluctuations

In 1995, prices for virgin PET hit record highs because of shortages of PET resins. Domestic prices for virgin PET resins increased from US$0.73 per pound at the end of 1994 to more than US$0.85 per pound by mid 1995. This increased prices of recycled PET to new record highs. These strong market conditions encouraged a number of new facilities to open that were capable of producing virgin PET.

At the same time that scrap reclaimers were getting settled into their current situations, prices entered a downslide, falling steadily during the first four months of 1996 before larger drops in April and May. This led to lows in August of US$0.05 per pound for delivered curbside bales and a FOB price of US$0.02-3 per pound. Deposit bottles receive a price premium of an additional US$0.02-3.5 per pound. The main cause of this decline was the increased virgin PET resin supply. North American and worldwide virgin PET production capacity flooded the plastics market in 1996 as new plants came on line and virgin resin in inventory had to be sold at distressed prices. With these low prices, many users of recycled PET switched to wide specification virgin PET, based on the economic advantages.

Effects on Industry

The price drop and the struggle to find markets for the recycled PET have forced many PET recyclers to close. With the excitement felt in the recycled PET industry in 1995, 10 new PET reclaimers burst onto the scene making the industry more than 30 strong. With the downturn in 1996, this was reversed and the industry is left with 8-12 reclaimers processing

PET. Of the approximately 25 reclaimers that have left the industry, only 20% were vertically integrated or involved in adding product value through compounding and alloying. The remaining reclaimers now have the ability to process and sell about 600 million pounds of recycled PET.

Vertical Integration

One of the characteristics which many of the remaining successful processors of PET feature is vertical integration into product manufacturing, and the use of compounding to add product value. The two largest PET reclaimers, Image Industries (Summersville, South Carolina) and Wellman Inc (Johnsonville, South Carolina), are both product integrated into the fibre market. In addition, Wellman, and its subsidiaries Day Products and Creative Forming, as well as PETCO Plastics (Montreal, Quebec) are producing PET sheet from their recycled PET. Products from polyester carpet to thermoformed sheet to T-shirts to car parts are all made by companies which manufacture products using the PET they have recovered. Approximately 290 million pounds of the 572 million pounds (51%) of PET recycled in 1996 was used for internal manufacturing of products. What all these companies have shown is that integration of some form can increase profits and decrease direct competition with virgin resin producers.

Advantages of Vertical Integration

The main lustre behind vertical integration into product manufacture is the economic advantages it offers. Current sale prices for recovered clear PET flake material are between US$0.31-35 per pound and between US$0.42-46 cents per pound for pelletised PET material. The costs for production of this recycled PET flake or pellets is US$0.7-15 cents less than the selling price, table 1. This represents a saving of between 25 35% over the purchase of material from another PET reclaimer. A lower raw material price translates to a saving in the price of the production, which, in turn, results in the ability to increase profits or undercut competitors' prices.

Table 1. : PET bottle recycling costs US$ per pound.

Granulated PET	
Raw material	2-6
Inbound freight	0-2
Processing Costs	13
Fixed costs	7
Total	22-28
Pelletised PET	
Raw material	2-6
Inbound freight	0-2
Processing costs	23
Fixed costs	10.5
Total	35.5-41.5

Besides the obvious economic advantages, product integration also allows usage of choice material, assurances of raw materials, and lower transportation costs. In addition, sale of a product removes some of the focus from the recycling operations, allowing survival even during downturns in the economics of PET recycling such as happened in 1996.

Quality Issues

Integration isn't the sole answer to success in the PET recycling industry. Companies such as Agra Industries (which included sheet production), Goodwill industries (purchased by RockTenn, which is a sheet producer), Plastic Recycling Alliance (sister company of Signode, which is a large PET strapping producer, both owned by ITW), and Nationwide Recyclers (sister company to Constar, which is a large PET bottle producer) were vertically integrated, but suffered problems with quality and production costs. This led each of these companies has since shut down.

Recycled PET Products

Some products that cane be manufactured from recycled PET include:

- Polyester fibre products e.g. carpets, sleeping bags, car boot linings, clothing, fibre fill and other textile applications
- Thermoformed sheet products e.g. laundry scoops, containers for fruit and berries, non-food containers and specialised food packaging
- Other applications e.g. hoses.

PET Alloying

In 1996, 20 million pounds of recycled PET was used in applications in which it was first combined with another resin. This process, known as alloying or compounding, greatly increases the chemical resistance, dimensional stability, and abrasion resistance. PET, both recycled and virgin, cannot easily be injection moulded. PET will randomly crystallise during normal thick wall moulding thus making the moulded part brittle and subject to warp. The process of adding an impact modifier to PET allows it to overcome those problems and allows it to be used in other applications.

Applications for these resins are in the industrial, lawn and garden, sports goods, electrical/electronic, and automotive areas. Product applications include:

- Car bumper bars, other panels and recreational vehicle parts
- Computer keyboards
- Boating equipment
- Protective helmets
- Lawn mower decks
- Tractor grilles
- Meter covers
- Conveyor components
- Consumer sporting goods

- Household appliances
- Telephone housings
- Industrial motor housings
- Pump components
- Lawn/garden machinery

PET/Polycarbonate Alloys

Alloys of PET and polycarbonate are common, using PET contents of 50-80%. Such alloys can be up to 25-30% less expensive than those produced using virgin material, typically acrylonitrile-butadiene-styrene (ABS). They can be produced in different colours and with or without flame retardants using processing techniques such as injection moulding, blow moulding, and sheet extrusion. PET/PC alloys are used for:

- Automotive Applications - automobile bumpers, grilles, wheel covers, and body panels, electrical and electronic components
- Power tools, and other appliances
- Consumer items - office machines and cartridges for copying machines.

Glass-Filled Compounds

Compounds have proven to be another growing market such as glass-filled compounds containing 35% glass fibre and minerals. This material is used in the automotive industry. For example, Ford uses it for grille opening reinforcing panels on its cars and trucks. Applications include:

- Headlamp brackets
- Exterior car door escutcheons
- Automotive window hardware
- Roof racks
- Pot handles
- Arms for chairs

All of these products require good dimensional stability, strength, stiffness, and paintability.

9

Introduction of Polylactic

Polylactic acid or Polylactide (PLA) is a type of bioplastic. Polylactic acid or polylactide (PLA) is a biodegradable, thermoplastic, aliphatic polyester derived from renewable resources, such as corn starch (in the U.S.) or sugarcanes (rest of world). Although PLA has been known for more than a century, it has only been of commercial interest in recent years, in light of its biodegradability.

PLA can be made from corn starch, or other starch-rich substances like maize, sugar or wheat. Packaging made from PLA is bio-degradable and reverts in less than 60 days in ideal conditions, namely in commercial composting installations.

However, PLA is still more expensive than many petroleum-derived commodity plastics, but its price has been falling as more production comes online. The degree to which the price will fall, and the degree to which PLA will be able to compete with non-sustainable petroleum-derived polymers, is uncertain.

Synthesis

Bacterial fermentation is used to produce lactic acid from corn starch or cane sugar. However, lactic acid cannot be directly polymerized to a useful product, because each polymerization reaction generates one molecule of water, the presence of which degrades the forming polymer chain to the point that only very low molecular weights are observed. Instead, lactic acid is oligomerized and then catalytically dimerized to make the cyclic lactide monomer. Although dimerization also generates water, it can be separated prior to polymerization. PLA of high molecular weight is produced from the lactide monomer by ring-opening polymerization using most commonly a stannous octoate catalyst, but for laboratory demonstrations tin(II) chloride is often employed. This mechanism does not generate additional water, and hence, a wide range of molecular weights are accessible.

Polymerization of a racemic mixture of L- and D-lactides usually leads to the synthesis of poly-DL-lactide (**PDLLA**) which is not crystalline but amorphous. Use of stereospecific catalysts can lead to heterotactic PLA which has been found to show crystallinity. The degree of crystallinity, and hence many important properties, is controlled by the ratio of D to L enantiomers used.

Chemical and physical properties

Due to the chiral nature of lactic acid, several distinct forms of polylactide exist: poly-L-lactide (**PLLA**) is the product resulting from polymerization of L,L-lactide (also known as L-lactide). PLLA has a crystallinity of around 37%, a glass transition temperature between 50-80 °C and a melting temperature between 173-178 °C.

Polylactic acid can be processed like most thermoplastics into fibre (for example using conventional melt spinning processes) and film. The melting temperature of PLLA can be increased 40-50 °C and its heat deflection temperature can be increased from approximately 60°C to up to 190 °C by physically blending the polymer with PDLA (poly-D-lactide). PDLA and PLLA form a highly regular stereo complex with increased crystallinity. The temperature stability is maximised when a 50:50 blend is used, but even at lower concentrations of 3-10% of PDLA, there is still a substantial improvement. In the latter case, PDLA acts as a nucleating agent, thereby increasing the crystallization rate. Biodegradation of PDLA is slower than for PLA due to the higher crystallinity of PDLA. PDLA has the useful property of being optically transparent.

Applications

Stereo complex blends of PDLA and PLLA have a wide range of applications, such as woven shirts (ironability), microwavable trays, hot-fill applications and even engineering plastics (in this case, the stereo complex is blended with a rubber-like polymer such as ABS). Such blends also have good form-stability and visual transparency, making them useful for low-end packaging applications. Progress in bio-technology has resulted in the development of commercial production of the D(-) form, something that was not possible until recently.

PLA is currently used in a number of biomedical applications, such as sutures, stents, dialysis media and drug delivery devices. It is also being evaluated as a material for tissue engineering. Because it is biodegradabie, it can also be employed in the preparation of bioplastic, useful for producing loose-fill packaging, compost bags, food packaging, and disposable tableware. In the form of fibres and non-woven textiles, PLA also has many potential uses, for example as upholstery, disposable garments, awnings, feminine hygiene products, and nappies.

PLA has been used as the hydrophobic block of amphiphilic synthetic block copolymers used to form the vesicle membrane of polymersomes.

PLA is a sustainable alternative to petrochemical-derived products, since the lactides from which it is ultimately produced can be derived from the fermentation of agricultural by-products such as corn starch or other carbohydrate-rich substances like maize, sugar or wheat.

PLA is more expensive than many petroleum-derived commodity plastics, but its price has been falling as production increases. The demand for corn is growing, both due to the use of corn for bioethanol and for corn-dependent commodities, including PLA.

Production

As of December 2005, Nature Works was the primary producer of PLA (bioplastic) in the United States.

Other companies involved in PLA manufacturing are Toyota (Japan), PURAC Biomaterials (The Netherlands), Hycail (The Netherlands), Galactic (Belgium), DURECT (US) and several Chinese manufacturers. The primary producer of PDLLA is PURAC, a wholly owned subsidiary of CSM located in the Netherlands.

Galactic and Total Petrochemicals operate a joint-venture, Futerro, that is developing a second generation of polylactic acid product. This project includes the building of a PLA pilot plant of 1500 tonnes/year in Belgium.

PLA has become a significant commercial bioplastic. Its clarity makes it useful for recyclable and biodegradable packaging, such as bottles, yogurt cups, and candy wrappers. It has also been used for food service ware, lawn and food waste bags, coatings for paper and cardboard, and fibres-for clothing, carpets, sheets and towels, and wall coverings. In biomedical applications, it is used for sutures, prosthetic materials, and materials for drug delivery.

PLA or polylactic acid is a polymer (chemical chains) made from the lactic acid that is a result of cornstarch fermentation. PLA polymers are used in plastics that are competitive with and often superior to hydrocarbon-based polymers and is completely renewable.

The U.S. produces annually between 70 and 80 billion pounds of plastics, mostly from "hydrocarbon" or petroleum-based sources. If we were to replace theses plastics with corn-based products, the process would consume 2 billion bushels of corn annually, representing a significant market for U.S. corn.

PLA competes against four main hydrocarbon-based resins—polyethylene, polystyrene, polypropylene and polyester. By comparison, PLA: is inherently clear, naturally glossy, resistant to moisture and grease and naturally stiff; can be processed using most conventional techniques and equipment and copolymerized with other materials; and it readily accepts fillers.

Maybe most important to early marketing efforts, PLA can be engineered to be biodegradable in controlled compost situations.

Only about 25 percent of the total market is appropriate for biodegradable plastics. Due to the significant market opportunity of both biodegradable and non-biodegradable plastics, companies continue to develop this market.

Most recently, Ingeo (PLA Fibre) was designated as a "new generic fibre" by the Federal Trade Commission, opening the doors to increased use.

Applications of Ingeo include clothing, wipes, carpet tiles, diapers, feminine hygiene products, upholstery, interior and outdoor furnishings, filtration and agricultural applications.

Polylactic Acid (PLA) Polymer Film

Polylactic acid (PLA) polymer film was degraded in abiotic and biotic environments to understand the role of microbes in the degradation process of lactic acid based polymers. The degradation studies were conducted in a well-characterized biotic system, an abiotic system, a sterile aqueous system, and a desiccated environment maintained at 40, 50, and 60 degrees C. The combination of experiments in different environments isolated the distinct effect of microbes, water, and temperature on the morphological changes in the polymer during degradation. Due to lack of availability of radiolabeled PLA, various analytical techniques were applied to observe changes in the rate and/or mechanism of degradation. CO2 evolved, weight loss, and molecular weights were measured to evaluate the extent of degradation. X-ray diffraction and differential scanning calorimetry techniques monitored the morphological changes in the polymer. FTIR was used as a semi quantitative tool to gather information about the chemistry of the degenerative process. Neither of the above analytical techniques indicated any difference in the rate or mechanism of degradation attributable to the presence of microorganisms. The extent of degradation increased at higher process temperatures. FTIR data were evaluated for significant statistical difference by t-test hypothesis. The results confirmed hydrolysis of ester linkage as the primary mechanism of degradation of PLA. On the basis of these data, a probable path of PLA degradation has been suggested.

Miscellaneous Facts

One of the little known facts about PLA is that it is quite an old material, discovered in the 1890s, but it has only now found a universal route to market in the form of biodegradable packaging.

PLA (Polylactic Acid) is an environmental friendly raw material substitute for traditional petroleum derived plastic products. It is a sustainable alternative as it is a product derived from agricultural by-products, such as cornstarch and other carbohydrate-rich substances - like Maize, sugar or wheat.

It is easier to process than the other available biodegradable materials, and has a wide range of applications.

Products made of PLA can be clear or opaque, flexible or rigid. They can be glossy, matt or tensile. PLA resists grease and oil, offers an excellent flavor/odor. This is why PLA is ideal for manufacturing packages and bags. All of our food products are certified safe for food contact. We are able to offer a complete selection of tableware, containers, cutlery, plates, bowls, and biodegradable plastic bags; all customizable for individual requirements.

In another application, PLA is being extruded into a high-performance synthetic fibre. Designers can use it to create new yarns, fabrics and garments. It is ideally suited for apparel, furnishing and nonwoven applications.

Plastarch Material (PSM)

Plastarch Material (PSM) is a biodegradable, thermoplastic resin. It is composed of starch combined with several other biodegradable materials. The starch is modified in order to obtain heat-resistant properties, making PSM one of few bioplastics capable of withstanding high temperatures. PSM began to be commercially available in 2005.

PSM is stable in the atmosphere, but biodegradable in compost, wet soil, fresh water, seawater, and activated sludge where microorganisms exist. It has a softening temperature of 257°F (125°C) and a melting temperature of 313°F (156°C).

It is also hygroscopic. The material has to be dried in a material dryer at 150°F (66°C) for five hours or 180°F (82°C) for three hours. For injection molding and extrusion the barrel temperatures should be at 340° +/- 10°F (171°C) with the nozzle/die at 360°F (182°C).

Due to how similar PSM is to other plastics (such as polypropylene and CPET), PSM can run on many existing thermoforming and injection molding lines. PSM is currently used for a wide variety of applications in the plastic market, such as food packaging and utensils, personal care items, plastic bags, temporary construction tubing, industrial foam packaging, industrial and agricultural film, window insulation, construction stakes, and horticulture planters.

Since PSM is derived from a renewable resource (corn), it has become an attractive alternative to petrochemical-derived products. Unlike plastic, PSM can also be disposed of through incineration, resulting in non-toxic smoke and a white residue which can be used as fertilizer.

Vinyon

Vinyon is a synthetic fibre made from polyvinyl chloride. In some countries other than the United States, vinyon fibres are referred to as **polyvinyl chloride fibres**. It can bind non-woven fibres and fabrics. It was invented in 1939.

It has the same health problems associated with chlorinated polymers. In the past, Vinyon was used a substitute for plant-based filters in tea bags.

Vinyon Fibre Characteristics

- doesn't flame, but softens at low temperatures (55 C)
- high resistance to chemicals
- Moisture absorption is less than 0.5% and moisture regained is less than 0.1%

Major Vinyon Fibre Uses

- industrial applications as a bonding agent for non-woven fabrics and products

Production

The U.S. Federal Trade Commission definition for vinyon fibre is "A manufactured fibre in which the fibre-forming substance is any long chain synthetic polymer composed of at least 85 percent by weight of vinyl chloride units (—CH2—CHCl—)."

First U.S. commercial vinyon fibre production: 1939, FMC Corporation, Fibre Division (formerly American Viscose).

Federal Trade Commission Definition for Vinyon Fibre: A manufactured fibre in which the fibre forming substance is any long-chain synthetic polymer composed of at least 85% weight of vinyl chloride units. (-CH_2 CHCl-$)_X$.

Characteristics and Vinyon Fibre Uses — Vinyon fibres soften at low temperatures but have high resistance to chemicals. They are most commonly used in industrial applications as a bonding agent for non-woven fabrics and products.

In some countries other than the United States, vinyon fibres are referred to as polyvinyl chloride fibres.

Aramid

Aramid fibres are a class of heat-resistant and strong synthetic fibres. They are used in aerospace and military applications, for ballistic rated body armor fabric, and as an asbestos substitute. The name is a shortened form of "aromatic polyamide". They are fibres in which the chain molecules are highly oriented along the fibre axis, so the strength of the chemical bond can be exploited.

History

Aromatic polyamides were first introduced in commercial applications in the early 1960s, with a meta-aramid fibre produced by DuPont under the tradename Nomex. This fibre, which handles similarly to normal textile apparel fibres, is characterized by its excellent resistance to heat, as it neither melts nor ignites in normal levels of oxygen. It is used extensively in the production of protective apparel, air filtration, thermal and electrical insulation as well as a substitute for asbestos. Meta-aramid is also produced in the Netherlands and Japan by Teijin under the tradename Teijinconex, in China by Yantai under the tradename New Star and a variant of meta-aramid in France by Kermel under the tradename Kermel.

Based on earlier research by Monsanto and Bayer, a fibre- para-aramid - with much higher tenacity and elastic modulus was also developed in the 1960s-1970s by DuPont and Akzo Nobel, both profiting from their knowledge of rayon, polyester and nylon processing.

Much work was done by Stephanie Kwolek in 1961 while working at DuPont, and that company was the first to introduce a *para-aramid* called Kevlar in 1973. A similar fibre called Twaron with roughly the same chemical structure was introduced by Akzo in 1978. Due to earlier patents on the production process, Akzo and DuPont had a patent war in the 1980s. Twaron is currently owned by the Teijin company (see Production).

Para-aramids are used in many high-tech applications, such as aerospace and military applications, for "bullet-proof" body armor fabric.

The Federal Trade Commission definition for aramid fibre is:

A manufactured fibre in which the fibre-forming substance is a long-chain synthetic polyamide in which at least 85% of the amide linkages, (-CO-NH-) are attached directly to two aromatic rings.

Production

World capacity of para-aramid production is estimated at about 41,000 tons/yr in 2002 and increases each year by 5-10%. In 2007 this means a total production capacity of around 55,000 tons/yr.

Polymer Preparation

Aramids are generally prepared by the reaction between an amine group and a carboxylic acid halide group. Simple AB homopolymers may look like:

$n\mathrm{NH_2\text{-}Ar\text{-}COCl}$ '! $\mathrm{-(NH\text{-}Ar\text{-}CO)_n\text{-}} + n\mathrm{HCl}$

The most well-known aramids (Nomex, Kevlar, Twaron and New Star) are AABB polymers. Nomex, New Star and

Teijinconex contain predominantly the meta-linkage and are poly-*metaphenylene isophtalamides* (MPIA). Kevlar and Twaron are both *p*-phenylene terephtalamides (PPTA), the simplest form of the AABB para-polyaramide. PPTA is a product of *p*-phenylene diamine (PPD) and terephtaloyl dichloride (TDC or TCl). Production of PPTA relies on a co-solvent with an ionic component (calcium chloride ($CaCl_2$)) to occupy the hydrogen bonds of the amide groups, and an organic component (N-methyl pyrrolidone (NMP)) to dissolve the aromatic polymer. Prior to the invention of this process by Leo Vollbracht, who worked at the Dutch chemical firm Akzo, no practical means of dissolving the polymer was known. The use of this system led to a patent war between Akzo and DuPont.

Spinning

After production of the polymer, the aramid fibre is produced by spinning the solved polymer to a solid fibre from a liquid chemical blend. Polymer solvent for spinning PPTA is generally 100% (water free) sulfuric acid (H_2SO_4).

Appearances

- Fibre
- Chopped fibre
- Powder
- Pulp

Other Types of Aramids

Besides meta-aramids like Nomex, other variations belong to the aramid fibre range. These are mainly of the **copolyamide** type, best known under the brand name Technora, as developed by Teijin and introduced in 1976. The manufacturing process of Technora reacts PPD and 3,4'-diaminodiphenylether (3,4'-ODA) with terephtaloyl chloride (TCl). This relatively simple process uses only one amide solvent and therefore spinning can be done directly after the polymer production.

Aramid Fibre Characteristics

Aramids share a high degree of orientation with other fibres such as Ultra high molecular weight polyethylene, a characteristic which dominates their properties.

General

- good resistance to abrasion
- good resistance to organic solvents
- nonconductive
- no melting point, degradation starts from 500°C
- low flammability
- good fabric integrity at elevated temperatures
- sensitive to acids and salts
- sensitive to ultraviolet radiation
- prone to static build-up unless finished

Para-aramids

- para-aramid fibres such as Kevlar and Twaron, provide outstanding strength-to-weight properties
- high Young's modulus
- high tenacity
- low creep
- low elongation at break (3.5%)
- difficult to dye - usually solution dyed [3]

Major Industrial Uses

- flame-resistant clothing
- heat protective clothing and helmets
- body armor, competing with PE based fibre products such as Dyneema and Spectra
- composite materials
- asbestos replacement (e.g. braking pads)

- hot air filtration fabrics
- tires, newly as Sulfron (sulfur modified Twaron)
- mechanical rubber goods reinforcement
- ropes and cables
- wicks for fire dancing
- optical fibre cable systems
- sail cloth (not necessarily racing boat sails)
- sporting goods
- drumheads
- wind instrument reeds, such as the Fibracell brand
- speaker woofers
- boathull material
- fibre reinforced concrete
- reinforced thermoplastic pipes
- tennis strings (e.g. by Ashaway and Prince tennis companies)
- hockey sticks (normally in composition with such materials as wood and carbon)

Para-aramid

- ***Kevlar***

Kevlar is the registered trademark for a light, strong para-aramid synthetic fibre, related to other aramids such as Nomex and Technora.

Developed at DuPont in 1965 by Stephanie Kwolek it was first commercially used in the early 1970s as a replacement for steel in racing tires. Typically it is spun into ropes or fabric sheets that can be used as such or as an ingredient in composite material components.

Currently, Kevlar has many applications, ranging from bicycle tires and racing sails to body armor because of its high strength-to-weight ratio—famously: "*...5 times stronger than steel on an equal weight basis...*"

A similar fibre called Twaron with roughly the same chemical structure was introduced by Akzo in 1978, and now manufactured by Teijin.

Properties

When Kevlar is spun, the resulting fibre has great tensile strength (ca. 3 620 MPa), and a relative density of 1.44. When used as a woven material, it is suitable for mooring lines and other underwater applications.

There are three grades of Kevlar: (i) Kevlar, (ii) Kevlar 29, and (iii) Kevlar 49. Typically, Kevlar is used as reinforcement in tires and rubber mechanical goods. Kevlar 29's industrial applications are as cables, in asbestos replacement, brake linings, and body armor. Kevlar 49 has the greatest tensile strength of all the aramids, and is used in plastic reinforcement for boat hulls, airplanes, and bicycles. The ultraviolet light component of sunlight degrades and decomposes Kevlar, a problem known as UV degradation, and so it is rarely used outdoors without protection against sunlight.

Production

Kevlar is synthesised in solution from the monomers 1,4-phenylene-diamine (*para*-phenylenediamine) and terephthaloyl chloride in a condensation reaction yielding hydrochloric acid as a byproduct. The result has liquid-crystalline behaviour, and mechanical drawing orients the polymer chains in the fibre's direction. Hexamet hylphosphoramide (HMPA) was the polymerization solvent first used, but toxicology tests demonstrated it provoked tumors in the noses of rats, so DuPont replaced it by a N-methyl-pyrrolidone and calcium chloride as the solvent. As this process was patented by Akzo (see above) in the production of Twaron, a patent war ensued.

Chemical Properties

Fibres of Kevlar consist of long molecular chains produced from PPTA (poly-paraphenylene terephthalamide). There are many inter-chain bonds making the material extremely strong. Kevlar derives part of its high strength from inter-molecular hydrogen bonds formed between the carbonyl groups and protons on neighboring polymer chains and the partial pi stacking of the benzenoid aromatic stacking interactions between stacked strands. These interactions have a greater influence on Kevlar than the van der Waals interactions and chain length that typically influence the properties of other synthetic polymers and fibres such as Dyneema. The presence of salts and certain other impurities, especially calcium, could interfere with the strand interactions and caution is used to avoid inclusion in its production. Kevlar's structure consists of relatively rigid molecules which tend to form mostly planar sheet-like structures rather like silk protein.

Thermal Properties

For a polymer, Kevlar has very good resistance to high temperatures, and maintains its strength and resilience down to cryogenic temperatures (-196°C); indeed, it is slightly stronger at low temperatures.

At higher temperatures the tensile strength is immediately reduced by about 10-20%, and after some hours the strength progressively reduces further. For example at 160°C about 10% reduction in strength occurs after 500 hours. At 260°C 50% strength reduction occurs after 70 hours.

At 450°C Kevlar sublimates.

Applications

Armor

Kevlar is well-known as a component of some bullet resistant vests and bullet resistant face masks. The PASGT helmet and vest used by United States military forces since the early 1980s both have Kevlar as a key component, as do their replacements. Other military uses include bulletproof facemasks used by sentries. Civilian applications include

Kevlar reinforced clothing for motorcycle riders to protect against abrasion injuries and also Emergency Service's protection gear if it involves high heat (e.g., tackling a fire), and Kevlar body armor such as vests for police officers, security, and SWAT.

Rope and cable

The fibre is used in woven rope and in cable, where the fibres are kept parallel within a polyethylene sleeve. Known as "Parafil", the cables have been used in small suspension bridges such as the bridge at Aberfeldy in Scotland. They have also been used to stabilise cracking concrete cooling towers by circumferential application followed by tensioning to close the cracks.

Sports equipment

It is used as an inner lining for some bicycle tires to prevent punctures, and due to its excellent heat resistance, is used for fire poi wicks. It is used for motorcycle safety clothing, especially in the areas featuring padding such as shoulders and elbows. It was also used as speed control patches for certain Soap Shoes models. In Kyudo or Japanese archery, it may be used as an alternative to more expensive hemp for bow strings. It is one of the main materials used for paraglider suspension lines.

Audio equipment

It has also been found to have useful acoustic properties for loudspeaker cones, specifically for bass and midrange drive units.

Electricity generation

Kevlar was used by scientists at Georgia Institute of Technology as a base textile for an experiment in electricity-producing clothing. This was done by weaving zinc oxide nanowires into the fabric. If successful, the new fabric would generate about 80 milliwatts per square meter.

Drumheads

Kevlar is sometimes used as a material in high tension drum heads usually used on marching snare drums. It allows for an extremely high amount of tension, resulting in a cleaner sound. There is usually some sort of resin poured onto the kevlar to make the head airtight, and a nylon top layer to provide a flat striking surface. This is one of the primary types of marching snare drum heads. Remo's falam slam patch is made with kevlar and is used to reinforce bass drum heads where the beater strikes.

Woodwind reeds

Kevlar is used in the woodwind reeds of Fibracell. The material of these reeds is a composite of aerospace materials designed to duplicate the way nature constructs cane reed. Very stiff but sound absorbing Kevlar fibres are suspended in a lightweight resin formulation.

Fibre Optic Cable

Kevlar is widely used as a protective outer sheath for fibre optic cable, as its strength protects the cable from damage and kinking.

Building construction

A retractable roof of over 60,000 square feet (5,575 square metres) of Kevlar was a key part of the design of Montreal's Olympic stadium for the 1976 Summer Olympics. It was spectacularly unsuccessful, as it was completed ten years late and replaced just ten years later in May 1998 after a series of problems.

Brakes

The chopped fibre has been used as a replacement for asbestos in brake pads. Dust produced from asbestos brakes is toxic, while aramids are a benign substitute.

Expansion joints and hoses

Kevlar can be found as a reinforcing layer in rubber bellows expansion joints and rubber hoses, for use in high temperature applications, and for its high strength. It is also found as a braid layer used on the outside of hose assemblies, to add protection against sharp objects.

Composite materials

Aramid fibres are widely used for reinforcing composite materials, often in combination with carbon fibre and glass fibre. The matrix for high performance composites is usually epoxy resin. Typical applications include monocoque bodies for F1 racing cars, helicopter rotor blades, tennis, table tennis, badminton and squash rackets, kayaks, cricket bats, and field hockey, ice hockey and lacrosse sticks.

- *Technora*

Technora is the brandname of Teijin for a aromatic copolyamid.

Production process

The manufacturing process of Technora reacts PPD and 3,4'-diaminodiphenylether (3,4'-ODA) with terephtaloy chloride (TCl). This relatively simple process uses only one amide solvent and therefore spinning can be done directly after the polymer production.

Major Industrial Uses

- Automotive and other industries:
    - Turbo hoses
    - high pressure hoses
    - Timing and V-belts
    - mechanical rubber goods reinforcement
- Linear Tension
    - Optical fibre cables (OFC)
    - ropes, wire ropes and cables

- o Umbilical cables
- o Electrical Mechanical Cable (EMC)
- Drumheads
- *Twaron*

Twaron is the brandname of Teijin Aramid for a para-aramid.

History

Twaron is a heat-resistant and strong synthetic fibre developed in the early 1970s by the Dutch company AKZO, division Enka, later Akzo Industrial Fibres. The research name of the para-aramid fibre was originally **Fibre X**, but it was soon called **Arenka**. Although the Dutch para-aramid fibre was developed only a little later than DuPont's Kevlar, introduction of Twaron as a commercial product came much later than Kevlar due to financial problems at the AKZO company in the 1970s.

A short overview of the history of Twaron:

- In 1960s start of research program for Fibre X.
- In 1972 the ENKA Research laboratory develops a para-aramid called **Arenka.**
- In 1973 Akzo decides to use sulfuric acid (H_2SO_4) as a solvent for spinning.
- In 1976 a pilot plant is built. First production in 1977.
- In 1984 the product is renamed **Twaron.**
- In 1986 commercial production is started at three locations and nine plants.
- In 1987 **Twaron** is introduced as a commercial product.
- In 1989 the aramid business of Akzo becomes an independent Business Unit called Twaron BV.
- Since 2000 Twaron BV is owned by the Teijin Group, now called Teijin Twaron BV and based in Arnhem, The Netherlands. The main production facilities for Twaron are in Emmen en Delfzijl.
- In 2007 Teijin Twaron expands for the fourth time in six years and also changes its name into Teijin Aramid.

Production

Polymer preparation

Twaron is a *p*-phenylene terephtalamide (P*p*PTA), the simplest form of the AABB para polyaramide. P*p*PTA is a product of *p*-phenylene diamine (PPD) and terephtaloy dichloride (TDC). To dissolve the aromatic polymer Twaron used a co-solvent of N-methyl pyrrolidone (NMP) and an ionic component (Calcium Chloride ($CaCl_2$) to occupy the hydrogen bonds of the amide groups. Prior to the invention of this process by Leo Vollbracht, working at the Dutch chemical firm AKZO, no practical means of dissolving the polymer was known. The use of this system by DuPont led to a patent war between AKZO and DuPont as Dupont initially used the carcinogenic HMPT (Hexamethylphosphoramide). Despite heavy research DuPont now also uses the AKZO patent to use the less hazardous NMP in the Kevlar process.

Spinning

After the production of the Twaron polymer in Delfzijl, the polymer is brought to Emmen, where fibres are produced by spinning the dissolved polymer into a solid fibre from a liquid chemical blend. Polymer solvent for spinning PPTA is generally 100% (water free) sulfuric acid (H_2SO_4). The polymer is dissolved by mixing frozen sulfuric acid in powder form with the polymer in powder form and gently heating the mixture. This process, which differs from the more difficult DuPont process, was invented by Henri Lammers and patented by AKZO.

Major Industrial Uses

Twaron is a para-aramid and is used in automotive, construction, sport, aerospace, military and industry applications, e.g., "bullet-proof" body armor, fabric, and as an asbestos substitute.

- Protective Gear (Heat Resistant / Ballistics):

- o flame-resistant clothing
- o protective clothing and helmets
- o Anti-cut or heat resistant gloves
- o sporting goods
- o fabrics
- o body armor
- Composites:
- o composite materials
- o technical paper
- o asbestos replacement
- o hot air filtration
- o sail cloth
- o speaker woofers
- o boat hull material
- o fibre reinforced concrete
- o drumheads
- Automotive:
- o brake pads
- o turbo hoses
- o V-belts and Timing belts
- o tires that incorporate Sulfron (sulfur modified Twaron)
- o mechanical rubber goods reinforcement
- Linear Tension:
- o optical fibre cables (OFC)
- o ropes, wire ropes, cables [3]
- o umbilical cables
- o Electrical Mechanical Cable (EMC)
- o Reinforced Thermoplastic Pipes
- *Vectran*

Vectran is a manufactured fibre, spun from a liquid crystal polymer created by Celanese Acetate LLC and now manufactured by Kuraray Co., Ltd. Chemically it is an aromatic polyester.

Properties

These fibres are noted for thermal stability at high temperatures, high strength and modulus, low creep, and good chemical stability. They are moisture resistant and are generally stable in hostile environments. They have gold color. They are often used in combination with some polyester as a coating around Vectran core; polyurethane coating can improve abrasion resistance and act as a water barrier. Vectran has melting point of 330 °C, with progressive strength loss from 220 °C. As it has high resistance to ultraviolet radiation, it can be used outside for long term, if inspected regularly.

Usage

They are used as reinforcing (matrix) fibres for ropes, cables, and advanced composite materials, professional bike tires, and in electronics applications. Perhaps most notably, Vectran is used as one of the five layers in NASA's current space suit design, and was the fabric used for the airbags on the Mars Pathfinder and twin Mars Exploration Rovers *Spirit* and *Opportunity* missions, which allowed those craft to "soft land" on Mars in 1997 and 2004, respectively. The material is expected to be used again on NASA's 2011 Mars Science Laboratory in the bridle cables. It is also used in the inflatable spacecraft developed by Bigelow Aerospace, the first of which (*Genesis I*) has been in space since July 12, 2006.

Production

Kuraray Co., Ltd. owns 100% of the worldwide Vectran production since 2005 as they acquired the Vectran business from Celanese Advanced Materials Inc. (CAMI), based in South Carolina, U.S.

In 2007 the total capacity of Vectran is about 600 tons/yr, but expanded to 1000 tons/yr in 2008.

Meta-aramid

- *Nomex*

Nomex (styled **NOMEX**) is a registered trademark for flame resistant meta-aramid material developed in the early 1960s by DuPont and first marketed in 1967.[1]

It can be considered an aromatic nylon, the *meta* variant of the *para*-aramid Kevlar. It is sold in both fibre and sheet forms and is used as a fabric wherever resistance from heat and flame is required. Nomex sheet is actually a calendered paper and made in a similar fashion. Nomex Type 410 paper is the original and one of the larger grade types made, mostly for electrical insulation purposes. The Spruance plant, in Richmond, VA, is the sole North American paper production site. The paper is used in electrical laminates such as circuit boards and transformer cores as well as fireproof honeycomb structures where it is saturated with a phenolic resin. Honeycomb structures such as these, as well as mylar-Nomex laminates are used extensively in aircraft construction. Both the firefighting and vehicle racing industries use Nomex to create clothing and equipment that can withstand intense heat. All aramids are heat and flame resistant but Kevlar, having a *para* orientation, can be molecularly aligned and gives high strength. *Meta* aramid cannot align during filament formation and has poor strength. Nomex fibre is made in the USA and in Spain (Asturias).

A Nomex hood is a common piece of Racing and fire fighting equipment. It is placed on the head on top of a fire fighter's face mask. The hood protects the portions of the head not covered by the helmet and face mask from the intense heat of the fire.

Race car drivers wear driving suits constructed of Nomex and or other fire retardant materials, along with Nomex gloves, long underwear, balaclavas, socks, and shoes to protect them in the event of a fire. The FIA and the SFI Foundation provide specifications for flame-resistant drivers clothing to be used in racing. The standards range from single layer suits that provide some protection against flash fires to much thicker multi layer SFI-15 suits required by the National Hot Rod Association that can protect a driver for up to 30 seconds against the intense heat generated by the nitromethane fuel they use.

Military pilots and aircrew wear flight suits made of over 92 percent Nomex to protect them from the possibility of cockpit fires and other mishaps. Recently, troops riding in ground vehicles have also begun wearing Nomex. The remaining material is typically Kevlar thread used to hold the fabric together at the seams.

In the U.S. space program, Nomex has been used for the Thermal Micrometeoroid Garment on the Extravehicular Mobility Unit (in conjunction with Kevlar and Gore-Tex) and ACES pressure suit, both for fire and extreme environment (water immersion to near vacuum) protection, and as thermal blankets on the payload bay doors, fuselage, and upper wing surfaces of the Space Shuttle Orbiter. It has also been used for the airbags for the Mars Pathfinder and Mars Exploration Rover missions, the Galileo atmospheric probe, the Cassini-Huygens Titan probe, and will be incorporated into the new Crew Exploration Vehicle that will replace the Shuttle after 2010.

Nomex has also been used for its unique acoustic qualities, the first time being used in Troy, NY at Rensselaer Polytechnic Institute's Experimental Media and Performing Arts Center EMPAC's main concert hall. Nomex reflects high-frequency sound, and increases mid- and low-level frequency sound. According to RPI President Shirley Ann Jackson, EMPAC is the first venue in the world to use Nomex for acoustic reasons.

The DuPont scientist responsible for discoveries leading to the creation of Nomex, Dr. Wilfred Sweeny, earned a DuPont Lavoisier Medal partly for this work in 2002.

Honeycomb-structured Nomex paper is also used as a spacer between layers of lead in the ATLAS Liquid Argon Calorimeter

- *Teijinconex*

Teijinconex is the brand name of Teijin for a meta-aramid fibre.

Teijin Conex® ("Conex") meta-aramid fibres are best known for their combination of heat resistance and strength. In addition, m-aramid fibres do not ignite, melt or drip, which is a major reason for their success in the fireproof apparel market. In comparison to commodity fibres, m-aramids offer better long-term retention of mechanical properties at elevated temperatures. M-aramids have a relatively soft hand and tend to process very similarly to commodity fibres, giving them a wide range of converted products.

Teijinconex is a meta-linked aromatic polyamide fibre known for its heat resistance. Teijin Conex is a synthetic organic fibre comprised of polymetaphenylene isophthalic amide, which is formed by the reaction from meta-phenylenediamine and isophthaloyl chloride.

Teijinconex does not melt in air, and offers excellent flame and combustion resistance. Teijinconex begins to decompose and carbonize at 400°C and has an LOI (Limiting Oxygen Index) of 30. It is a white, highly functional fibre that can be used from clothing to industrial materials."

The following properties characterize Conex:

- Resistant to high temperatures
- Inherently flame retardant
- Can be used in a wide range of temperatures

10

Special Types of Synthetic Textile

Saran (Plastic)

Saran is the trade name for a number of polymers made from vinylidene chloride (especially polyvinylidene chloride or PVDC), along with other monomers. The main advantage of Saran film is a very low permeability to water vapor, flavor and aroma molecules, and oxygen compared to other plastics. The barrier to oxygen prevents food spoilage, and the barrier to flavor and aroma molecules helps food retain its flavor and aroma.

Saran is best known for having been used in *Saran Wrap*—a thin, clingy plastic wrap that was sold in rolls and used primarily for wrapping food. Saran Wrap is now a brand of S. C. Johnson & Son. Environmental concerns with halogenated materials have led to the removal of PVDC from the Saran Wrap product which is now composed of polyethylene. With this change, Saran Wrap no longer provides the excellent oxygen barrier and therefore the food spoilage prevention benefit of the original product formulation.

In 1942, fused layers of original-specification Saran or PVDC were used to make ventilating insoles for boots, later adopted by the U.S. Army for use in their Jungle boot. PVDC is also used for high-quality doll hair that is valued by collectors for its shine, softness, and its ability to retain its style and curl.

In some jurisdictions, the name Saran is a registered trademark of the Dow Chemical Company, while in others, it has lost trademark status and become a generic term for these polymers. In Japan, Saran wrap is a trademark of Asahi Kasei.

History

Saran was first inadvertently discovered at Dow Chemical in 1933 when a lab worker, Ralph Wiley was having trouble washing beakers used in his process of developing a dry cleaning product. It was initially developed into a spray that was used on US fighter planes, and later automobile upholstery to protect them from the elements. Dow Chemical later refined Saran, eliminating its green hue and ridding it of its offensive odor. In 1949 Saran Wrap was invented and quickly became a common way to store food and keep it fresh.

Spandex

Spandex or **elastane** is a synthetic fibre known for its exceptional elasticity. It is stronger and more durable than rubber, its major non-synthetic competitor. It was invented in 1959 by DuPont chemist Joseph Shivers. When first introduced it revolutionized many areas of the clothing industry.

"Spandex" is a generic name and not derived from the chemical name of the fibre, which most manufactured fibres are, but an extension of the word *expand*. "Spandex" is the preferred name in North America; elsewhere it is referred to as "elastane".

The most famous brand name associated with spandex is *Lycra*, a trademark of Invista (formerly part of DuPont). Such is the prominence of the Lycra brand that it has become a genericised trademark in many parts of the world, used to describe any kind of spandex. Invista discourages such use, protecting its trademark vigorously.

Other spandex trademarks include **Elaspan** (also Invista's), **ROICA & Dorlastan** (Asahi Kasei) and **Linel** (Fillattice).

Spandex Fibre Production

Spandex fibres are produced in four different ways including melt extrusion, reaction spinning, solution dry spinning, and solution wet spinning. All of these methods include the initial step of reacting monomers to produce a prepolymer. Once the prepolymer is formed it is reacted further in various ways and drawn out to produce a long fibre. The solution dry spinning method is used to produce over 90% of the world's spandex fibres.

Solution Dry Spinning

Step 1: The first step is to produce the prepolymer. This is done by mixing a macroglycol with a diisocyanate monomer. The two compounds are mixed together in a reaction vessel to produce a prepolymer. A typical ratio of glycol to diisocyanate is 1:2.

Step 2: The prepolymer is further reacted with an equal amount of diamine. This reaction is known as *chain extension reaction*. The resulting solution is diluted with a solvent to produce the spinning solution. The solvent helps make the solution thinner and more easily handled and then can be pumped into the fibre production cell.

Step 3: The spinning solution is pumped into a cylindrical spinning cell where it is cured and converted into fibres. In this cell, the polymer solution is forced through a metal plate called a spinneret. This causes the solution to be aligned in strands of liquid polymer. As the strands pass through the cell, they are heated in the presence of a nitrogen and solvent gas. This process causes the liquid polymer to chemically react and form solid strands.

Step 4: As the fibres exit the cell, an amount of solid strands are bundled together to produce the desired thickness. Each fibre of spandex is made up of many smaller individual fibres that adhere to one another due to the natural stickiness of their surface.

Step 5: The resulting fibres are then treated with a finishing agent. This can be magnesium stearate or another polymer. This process prevents the fibres sticking together and aids in textile manufacture. The fibres are then transferred through a series of rollers onto a spool.

Step 6: When the spools are filled with fibre, they are put into final packaging and shipped to textile manufacturers.

Major Spandex Fibre Uses

- Apparel and clothing articles where stretch is desired, generally for comfort and fit, such as:
  - athletic, aerobic, and exercise apparel
  - wetsuits
  - swimsuits/bathing suits
  - competitive swimwear
  - netball body suits
  - brassiere straps and bra side panels
  - ski pants
  - disco jeans
  - slacks
  - hosiery
  - leggings
  - socks
  - diapers
  - skinny jeans

- o belts
- o underwear
- o dance belts worn by male ballet dancers and others
- • Compression garments such as:
- o surgical hose
- o support hose
- o cycling shorts
- o wrestling singlet
- o one piece rowing suits
- o foundation garments
- o motion capture suits
- • Shaped garments such as bra cups
- • Home furnishings, such as microbead pillows

In clothing it usually appears as a small percentage of total material. In North America it is rare in men's cheaper clothing, but prevalent in women's. It is used more often in women's as their clothes are usually more form-fitting. It is usually mixed with a greater percentage of one other textile such as cotton, polyester, or others. This keeps the reflection of light reduced to being hardly noticeable.

Spandex in Popular Culture

Superheroes :In comic books, superheroes and superheroines commonly wear costumes thought to be made of spandex. However, early superhero comics predate the invention of spandex (Superman-1938, Batman-1939, Captain America-1941). Printing processes for early color comics only rendered images with distinctly separate solid blocks of color well. Overprinting and color mixing yielded inconsistent results and bad looking muddy colors.

Because spandex is skintight, as many superhero costumes appear to be drawn, and because spandex is almost exclusively made in the same bright solid colors as the early Golden Age comics, the after-the-fact assumption of spandex composition was made. The same assumption of costume composition is also made for latex / rubber garments, which are also solid in colour and skin-tight.

In Japan, spandex is the common material for costumes used in the popular Super Sentai series (known overseas as *Power Rangers*). The first use of the material was in 1983's *Kagaku Sentai Dynaman*.

1970s/'80s rock/metal : During the 1970s and 1980s, spandex leggings rose in popularity amongst many rock and heavy metal bands, particularly British NWOBHM and American glam metal bands. The main reasons for this massive, almost universal, embracement of spandex amongst rock/metal bands was because spandex retained its stretchy, tight fitting quality, even after extended wear. Denim jeans and leather strides tended to sag and wear, while spandex did not. Also, the stretchiness of the material did not constrict musicians' movement onstage, allowing them to perform high kicks, or to rest their feet on monitors. Some of the rock/metal bands who used spandex leggings included Queen, Ratt, Van Halen, Mötley Crüe, Judas Priest, Scorpions, Iron Maiden, Saxon, Whitesnake, Bon Jovi and Twisted Sister. By the end of the 1980s and the decline of glam metal, and metal in general, with the advent of grunge, spandex fell out of fashion and many older glam bands found themselves being referred to as 'Spandex Jockeys'.

1970s/80s country : While glam metal bands were getting into the spandex craze, so were many glam-oriented Country stars, especially women like Dolly Parton, Margo Smith, and Dottie West. Dottie West is probably the best-known out of any Country singer for wearing spandex outfits on stage.

Spandex fetishism is a fetishistic attraction to people wearing stretch fabrics or, in certain cases, to the garments themselves. Spandex garments are often worn by swimmers, dancers, contortionists and circus performers, and spandex fetishists may incorporate fantasies about these activities into their fetish.

One reason why spandex and other tight fabrics may be fetishised is that the garment forms a "second skin," acting as a fetishistic surrogate for the wearer's own skin. Wearers of skin-tight nylon and cotton spandex garments can appear naked or coated in a shiny or matte substance like paint. The tightness of the garments may also be seen as sexual bondage. Another reason is that nylon-spandex fabric (preferred by many spandex fetishists) is often produced with a very smooth and silk-like finish, which lends a tactile dimension to the fetish - as well as a visual one.

In comic books, superheroes, superheroines, and supervillains are generally depicted as wearing costumes made of spandex or a similar material. The skintight, very flashy and bright-coloured costumes that often cover only just enough of the body to be presentable. Fantasies involving superheroes or the wearing of superhero costumes are commonly associated with spandex fetishism.

Full-body suits called zentai entirely immerse the wearer in skin tight fabric. The suits are essentially catsuits with gloves, feet, and a hood. The wearer gets to experience total enclosure and those who enjoy erotic objectification might make use of the garment's anonymizing aspect. Zentai fetishism appears to be quite popular in Japan and Europe. The word *zentai* means *whole body* in Japanese.

Pantyhose fetishism can be viewed as a class of spandex fetishism.

Vinalon

Vinalon is a synthetic fibre, produced from polyvinyl alcohol using anthracite and limestone as raw materials. Vinalon was first developed by the Korean scientist Ri Sung Gi at the Takatsuki chemical research institute in 1939. The fibre was largely ignored until Ri defected to North Korea in 1950. Trial production began in 1954 and in 1961 the massive February 8 Vinalon Complex was built in Hamhung. Its success and widespread usage in North Korea is often pointed to in propaganda as an example of the success of the *juche* philosophy. Hamhung remains a major production centre for vinalon; in 1998, a vinalon factory opened up in South Pyongan.

The factory complex also produces other chemicals. Some North Korean defectors have claimed that Dr. Lee Seung-ki was involved in chemical weapons research and that the complex was used to produce them.

Vinalon, also known as Juche fibre, has become the national fibre of North Korea and is used for the majority of textiles, outstripping fibre such as cotton or Nylon, which are only produced in small amounts in North Korea. Other than clothing, Vinalon is also used for canvas shoes, ropes and quilt wadding.

Vinalon is resistant to heat and chemicals but has numerous disadvantages: it is stiff, uncomfortable, shiny, prone to shrinking and difficult to dye. It is not produced outside of North Korea.

Modal

Modal is a bio-based fibre made by spinning reconstituted cellulose from beech trees. It is about 50% more hygroscopic, or water-absorbent, per unit volume than cotton is. It is designed to dye just like cotton, and is color-fast when washed in warm water. Modal is essentially a variety of rayon.

Textiles made from Modal are resistant to shrinkage and fading. They are smooth and soft, more so than even mercerized cotton, to the point where mineral deposits from hard water, such as lime, do not stick to the fabric surface. Like pure cotton, modal should ideally be ironed after washing.

Lenzing Modal is a registered trademark of Lenzing AG, an Austrian company specializing in textiles and fibres, particularly natural fibres made from cellulose. In the United States, pure Modal has begun being used in household linens such as towels, bathrobes, and bed sheets, and the fabric has experienced increased popularity in recent years.

Indian textile companies, in particular, have taken to Modal, and were expected to produce around 4000 tons of it in 2005.

Modal can be found in some types of socks. Those in the men's department of British Home Stores contain modal. Merona branded men's underwear was sold at Target stores in the United States in 2007. They consisted of 92% Modal and 8% Spandex and were manufactured in Malaysia. Jockey No Panty Line Promise® women's line also contains 92% modal/ 8% spandex. Old Navy, Banana Republic, Gap Body, Topo Ranch,Victoria's Secret, and G-Star Raw's Correct Line have all begun to blend the fibre with fabrics in several of their clothing products.

Hanky Panky, a major lingerie company, also uses Modal in their underwear. Three Dots, a t-shirt company, has been using modal blends, such as Supima Modal and Cotton Modal, in their t-shirts as well.

Polybenzimidazole or PBI

Polybenzimidazole or **PBI** fibre (1983) is a synthetic fibre with an extremely high melting point that also does not readily ignite, because of its exceptional thermal and chemical stability. The Federal Trade Commission definition for PBI fibre is "A manufactured fibre in which the fibre-forming substance is a long chain aromatic polymer having recurring imidazole groups as an integral part of the polymer chain." Kermel, a polyamidoimide, is a related "high performance polymer."

PBI is used to fabricate high-performance protective apparel such as fire fighter turnout coats and suits, astronaut space suits, high temperature protective gloves, welders' apparel, race driver suits, braided packings, and aircraft wall fabrics.

Production

PBI is prepared by step-growth polymerization from 3,3',4,4'-tetraaminobipheny and dipheny isophthalate (an ester of Isophthalic acid). The resulting polymer is processed as a solution in dimethyl acetamide via a dry spinning process. The first U.S. commercial PBI fibre was produced by Celanese.

Polyphenylene sulfide ("PPS") is an organic polymer consisting of aromatic rings linked with sulfides. Synthetic fibre and textiles derived from this polymer are known to resist chemical and thermal attack. PPS is used to make filter fabric for coal boilers, papermaking felts, electrical insulation, specialty membranes, gaskets, and packings. PPS is the precursor to a conducting polymer of the semi-flexible rod polymer family. The PPS, which is otherwise insulating, can be converted to the semiconducting form by oxidation or use of dopants.

Chevron Phillips manufactures the most common form in industry under the brand name Ryton.

An easy way to identify the plastic is by the metallic sound it makes when struck.

Fibre Characteristics

PPS is one of the most important high temperature polymers because it exhibits a number of desirable properties. These properties include resistance to heat, acids and alkalies, and to mildew, to bleaches, aging, sunlight, and abrasion. It absorbs only small amounts of solvents and resists dyeing.

Production

The Federal Trade Commission definition for sulfar fibre is "A manufactured fibre in which the fibre-forming substance is a long chain synthetic polysulfide in which at least 85% of the sulfide (—S—) linkages are attached directly to two (2) aromatic rings."

The PPS (polyphenylene sulfide) polymer is formed by reaction of sulfide with p-dichlorobenzene:

$$ClC_6H_4Cl + Na_2S \rightarrow 1/n\ [C_6H_4S]_n + 2\ NaCl$$

The process for commercially producing PPS (Ryton) was developed by Dr. H. Wayne Hill Jr. and Mr. James T. Edmonds at Phillips Petroleum Company. N-methylpyrrolidone is used as the reaction solvent since it is stable at the high temperatures required for the synthesis and it dissolves both the sulfiding agent and the oligomeric intermediates. The major challenge is controlling the degree of polymerization, which is in part controlled by managing the water content of the reaction mixture. The polymer is extruded by melt spinning to produce both staple and filament fibres.

The first U.S. Commercial Sulfar Fibre was produced in 1983 by Phillips Fibres Corporation, a subsidiary of Phillips 66 Company. Approximately 46M kg were produced in 1995 with a projected growth of 11% per year.

PPS is marketed by different brand names and companies. Examples of these are: Fortron by Ticona, as a linear polymer, Ryton PPS by Chevron Phillips, as cross-linked type of polymer, and Sulfar by Testori.

Lyocell

Lyocell is a fibre made from wood pulp cellulose. It was first manufactured in 1987 by Courtaulds Fibres UK at their pilot plant S25. The only current manufacturer in the United States is Lenzing Inc, who market it under the trademarked brand name **Tencel**. There is another larger plant with over twice as much capacity at Grimsby in the United Kingdom.

The Federal Trade Commission defines Lyocell as "a cellulose fabric that is obtained by an organic solvent spinning process". It classifies the fibre as a sub-category of rayon.

The fibre is used in the production of many clothes, such as jeans, trousers and coats.

Major properties

Lyocell was introduced to consumers in 1991 and

originally marketed as a type of Rayon. The only current manufacturer in the United States is Lenzing Inc, who market it under the trademarked brand name Tencel. Lyocell is produced from wood pulp cellulose. Lyocell shares many properties with other cellulosic fibres such as cotton, linen, ramie and rayon. Some main characteristics of lyocell fibres are that it is soft, absorbent, very strong when wet or dry, and resistant to wrinkles; it can be machine- or hand-washed or drycleaned, it drapes well, and it can be dyed many colors, as well as simulating a variety of textures like suede, leather, or silk. Fibre Source

Production Method

Lyocell is created through a process called solvent spinning. The wood pulp is dissolved in N-Methylmorpholine N-oxide, creating a solution called "dope." The dope is then pushed through a spinneret to form the individual fibres. After the dope has been spun into lyocell fibres, it is washed and the chemicals are retrieved from the water, purified, and recycled. (Kadolph, Sara, and Anna Langford. *Textiles:Ninth Edition*. Prentice Hall. 2002.) Since there is little by product, this process is relatively eco-friendly.

Uses

Currently Lyocell is more expensive to produce than cotton or rayon, but is included in many everyday items. Staple fibres are used in apparel items such as denim, chino, underwear and other casual wear clothing, even in bath towels. Filament fibres are used in items that have a silkier appearance such as women's clothing and men's dress shirts. Lyocell can be blended with a variety of other fibres such as silk, cotton, rayon, polyester, linen, nylon, and wool. Lyocell is also used in conveyor belts, specialty papers and medical dressings.

Ultra high molecular weight polyethylene (UHMWPE), also **known** as **high-modulus polyethylene (HMPE)** or **high-**

performance polyethylene (HPPE), is a subset of the thermoplastic polyethylene. It has extremely long chains, with molecular weight numbering in the millions, usually between 2 and 6 million. The longer chain serves to transfer load more effectively to the polymer backbone by strengthening intermolecular interactions. This results in a very tough material, with the highest impact strength of any thermoplastic presently made. It is highly resistant to corrosive chemicals, with exception of oxidizing acids. It has extremely low moisture absorption, has a very low coefficient of friction, is self-lubricating, and is highly resistant to abrasion (15 times more resistant to abrasion than carbon steel). Its coefficient of friction is significantly lower than that of nylon and acetal, and is comparable to that of Teflon, but UHMWPE has better abrasion resistance than Teflon. It is odorless, tasteless, and nontoxic.

Polymerisation of UHMWPE was commercialised in the 1950s by Ruhrchemie AG, which changed names over the years; today UHMWPE powder materials are produced by Ticona. UHMWPE is available commercially either as consolidated forms, such as sheets or rods, and as fibres. UHMWPE powder may also be directly molded into the final shape of a product. Because of its resistance to wear and impact, UHMWPE continues to find increasing industrial applications, including the automotive and bottling sectors, for example. Since the 1960s, UHMWPE has also been the material of choice for total joint arthroplasty in orthopedic and spine implants.

UHMWPE fibres, commercialised in the late 1970s by the Dutch chemicals company DSM, are widely used in ballistic protection, defense applications, and increasingly in medical devices as well.

Structure and Properties

UHMWPE is a type of polyolefin, and, despite relatively

weak Van der Waals bonds between its molecules, derives ample strength from the length of each individual molecule. It is made up of extremely long chains of polyethylene, which all align in the same direction. Each chain is bonded to the others with so many Van der Waals bonds that the whole can support great tensile loads.

When formed to fibres, the polymer chains can attain a parallel orientation greater than 95% and a level of crystallinity of up to 85%. In contrast, Kevlar derives its strength from strong bonding between relatively short molecules.

The weak bonding between olefin molecules allows local thermal excitations to disrupt the crystalline order of a given chain piece-by-piece, giving it much poorer heat resistance than other high-strength fibres. Its melting point is around 144 to 152 degrees Celsius, and, according to DSM, it is not advisable to use UHMWPE fibres at temperatures exceeding 80 to 100 °C for long periods of time. It becomes brittle at temperatures below -150 °C.

The simple structure of the molecule also gives rise to surface and chemical properties that are rare in high-performance polymers. For example, the polar groups in most polymers easily bond to water. Because olefins have no such groups, UHMWPE does not absorb water readily, but it also does not get wet easily, which makes bonding it to other polymers difficult. For the same reasons, skin does not interact with it strongly, making the UHMWPE fibre surface feel slippery. In a similar manner, aromatic polymers are often susceptible to aromatic solvents due to aromatic stacking interactions, an effect aliphatic polymers like UHMWPE are immune to. Since UHMWPE does not contain chemical groups (such as esters, amides or hydroxylic groups) that are susceptible to attack from aggressive agents, it is very resistant to water, moisture, most chemicals, UV radiation, and micro-organisms.

Under tensile load, UHMWPE will deform continually as long as the stress is present - an effect called *creep*.

Annealing : To anneal UHMWPE, the material should be heated to 135 °C to 138 °C in an oven or a liquid bath of silicone oil or glycerine. The material must then be cooled down at a rate of 5 °C/h to at least 65 °C. Finally, the material should be wrapped in an insulating blanket for 24 hours to bring to room temperature.

Production : UHMWPE is synthesized from monomers of ethylene, which are bonded together to form what is called *ultra-high-molecular-weight polyethylene* (or UHMWPE). These are molecules of polyethylene that are several orders of magnitude longer than familiar high-density polyethylene due to a synthesis process based on metallocene catalysts. In general, HDPE molecules have between 700 and 1,800 monomer units per molecule, whereas UHMWPE molecules tend to have 100,000 to 250,000 monomers each.

Finished UHMWPE is produced by 4 major methods: compression molding, ram extrusion, gel spinning, sintering, and kneading.

In gel spinning, a precisely-heated gel of UHMWPE is processed by an extruder through a spinneret. The extrudute is drawn through the air and then cooled in a water bath. The end-result is a fibre with a high degree of molecular orientation, and therefore exceptional tensile strength. Gel spinning depends on isolating individual chain molecules in the solvent so that intermolecular entanglements are minimal. Entanglements make chain orientation more difficult, and lower the strength of the final product.

Trade names : *Dyneema* is a registered trademark of Royal DSM (The Netherlands). Dyneema was invented by DSM in 1979. It has been in commercial production since 1990 at a plant in Heerlen, the Netherlands. In the Far East, DSM has a cooperation agreement with Toyobo for commercial production in Japan. In the United States, DSM has a production facility in Greenville, North Carolina, which is the largest production facility in the United States for UHMWPE fibre.

Honeywell developed a product identical in chemical structure, which is sold under the brand name *Spectra*. Though the production details are somewhat different, the resulting materials are comparable in properties.

Other trade names for powdered or consolidated UHMWPE materials include *TIVAR* by Quadrant EPP, *Polystone-M* by Röchling Engineering Plastics, *Tensylon* by Integrated Textile Systems, *GARDUR* by Garland Manufacturing, and *GUR* by Ticona.

Usage

Fibre applications : Ballistic vests are made of UHMWPE.

Dyneema and Spectra

Dyneema and **Spectra** are gel spun through a spinneret to form oriented-strand synthetic fibres of UHMWPE, which have yield strengths as high as 2.4 GPa and density as low as 0.97 kg/l (for Dyneema SK75). High strength steels have comparable yield strengths, and low carbon steels have yield strengths much lower (around 0.5 GPa). Since steel has a density approximately equal to 7.8 kg/l, this gives strength/weight ratios for these materials in a range from 10 to 100 times higher than for steel. Strength to weight ratios for Dyneema are about 40% higher than for Aramid.

UHMWPE fibres are used in ballistic vests, ballistic helmets and ballistic vehicle protection, cut resistant gloves, bow strings, climbing equipment, fishing line, spear lines for spearguns, high-performance sails, suspension lines on sport parachutes and paragliders, rigging in yachting, kites, and kites lines for kites sports. Spectra is also used as a high-end Wake board line.

For body armor, the fibres are, in general, aligned and bonded into sheets, which are then layered at various angles to give the resulting composite material strength in all directions. Recently-developed additions to the US Military's Interceptor body armor, designed to offer arm and leg protection, are said to utilise a form of Spectra or Dyneema fabric.

Spun UHMWPE fibres excel as fishing line, as they have less stretch, are more abrasion-resistant, and are thinner than traditional monofilament line.

In climbing, cord and webbing made of combinations of UHMWPE and nylon yarns have gained popularity for their low weight and bulk, though, unlike their nylon counterparts, they exhibit very low elasticity, making them unsuitable for limiting forces in a fall. Also, low elasticity translates to low toughness. The fibre's very high lubricity leads to poor knot-holding ability, and has led to the recommendation to use the Triple fisherman's knot rather than the traditional double fisherman's knot in 6mm UHMWPE *core* cord to avoid a particular failure mechanism of the double fisherman's, where first the sheath fails at the knot, then the core slips through.

Owing to its low density, ships' hawsers and cables can be made from the fibre, and float on sea water.

It is used in skis and snow boards, often in combination with carbon fibre, reinforcing the fibreglass composite material, adding stiffness and improving its flex characteristics. The UHMWPE is often used as the base layer, which contacts the snow, and includes abrasives to absorb and retain wax.

High-performance ropes for sailing and parasailing are made of UHMWPE, due to their low stretch and high strength.

Dyneema was used for the 30-kilometre space tether in the ESA/Russian Young Engineers' Satellite 2 of September, 2007.

The extremely low friction coefficient of UHMWPE makes it a common top sheet for boxes in terrain parks.[1]

Medical Applications in Total Joint Replacement : UHMWPE has over 40 years of clinical history as a successful

biomaterial for use in hip, knee, and most recently (since the 1980s), for spine implants. An online repository of information and review articles related to medical grade UHMWPE, known as the **UHMWPE Lexicon**, was started online in 2000.

Joint replacement components have historically been made from "GUR" resins. These powder materials are produced by Ticona, typically converted into semi-forms by companies such as Quadrant and Orthoplastics [12], and then machined into implant components and sterilised by device manufacturers.

UHMWPE was first used clinically in 1962 by Sir John Charnley and emerged as the dominant bearing material for total hip and knee replacements in the 1970s. Details about the "discovery" of UHMWPE for orthopedic applications by Charnley and his engineering associate Harry Craven are available. Throughout its history, there were unsuccessful attempts to modify UHMWPE to improve its clinical performance until the development of highly-crosslinked UHMWPE in the late 1990s.

One unsuccessful attempt to modify UHMWPE was by blending the powder with carbon fibres. This reinforced UHMWPE was released clinically as "Poly Two" by Zimmer in the 1970s. The carbon fibres had poor compatibility with the UHMWPE matrix and its clinical performance was inferior to virgin UHMWPE.

A second attempt to modify UHMWPE was by high pressure recrystallisation. This recrystallised UHMWPE was released clinically as "Hylamer" by DePuy in the late 1980s. When gamma irradiated in air, this material exhibited susceptibility to oxidation, resulting in inferior clinical performance related to virgin UHMWPE. Today, the poor clinical history of Hylamer is largely attributed to its sterilisation method, and there has been a resurgence of interest in studying this material (at least among certain research circles). Hylamer fell out of favor in the United States in the late 1990s with the development of highly-crosslinked UHMWPE materials, however negative clinical reports from Europe about Hylamer continue to surface in the literature.

Highly-crosslinked UHMWPE materials were clinically introduced starting in 1998 and have rapidly become the standard of care for total hip replacements, at least in the United States. These new materials are cross linked with gamma or electron beam radiation (50-105 kGy) and then thermally processed to improve its oxidation resistance. Five-year clinical data, from several centers, are now available demonstrating its superiority relative to conventional UHMWPE for total hip replacement. Clinical studies are still underway to investigate the performance of highly cross linked UHMWPE for knee replacement.

Another important medical advancement for UHMWPE in the past decade has been the increase in use of fibres for sutures. Medical grade fibres for surgical applications are produced by DSM under the "Dyneema Purity" trade name.

Skis and snow boards : The bottom of most modern skis— the surface that contacts the snow — is coated with UHMWPE, treated for compatibility with waxes and with epoxy base material. These treated materials are known as **P-tex, Isospeed**, or **Durasurf**. Because the material is a thermoplastic, gouges can easily be filled.

Small Bicycle Parts : Free style BMX company Tree Bicycle Co. uses UHMWPE for bicycle front sprocket guards to protect them during a stunt in which a rider grinds on their sprocket guard.

Skating : Aggressive in line skating use UHMWPE for sole plates and frames to assist how well the skates slide when performing grinding tricks

Garland Manufacturing has sponsored both solar cars and in line skating projects with its GARDUR brand of UHMWPE.

Most high-level short track speed skaters use a cut-resistant suit made of Dyneema or a comparable textile like Kevlar.

Skateboarding : Downhill and trick slide skateboarders use UHMWPE for the plastic pucks used on sliding gloves.

Hockey : Some ice hockey and field hockey sticks use Dyneema (US patent 5333857) as reinforcing material for the body of the stick. A sheet of the material can also be used as artificial ice for dry-land hockey stick handling practice.

Rock climbing : Rock climbers may use a sling made of dyneema because of its high tensile strength to weight ratio, and rarer Prusiks composed of dyneema can also be found, though their lower melting point versus nylon makes them largely unsuitable for this job, resulting in the creation of nylon sheathed Prusiks with dyneema cores.

Kite flying : Dyneema and its sister line Spectra have superseded Kevlar line sets used for dual and quad line steerable (stunt) kites. These UHMWPE lines have low stretch, low weight and a great strength to diameter ratio. They are generally not used for single line kites as Dyneema and Spectra are very slippery and it is difficult for even a gloved hand to grip the line. On a crowded flying field Dyneema lines are subject to being severed by thin Dacron lines due to melting from friction. Dyneema and Spectra have relatively low melting points of 144 - 152 deg C. The world kite altitude record for a single kite is being attempted in Australia using 11 km of Dyneema line.

Manufacturing : UHMW is used in the manufacture of PVC(vinyl) windows and doors, as it can stand up to the heat required to soften the PVC based materials and is used as a form/chamber filler for the various PVC shape profiles in order for those materials to be 'bent' or shaped around a template.

Zylon

Zylon is a trademarked name for a range of thermoset polyurethane synthetic polymer materials manufactured by the Toyobo Corporation. Zylon was invented and developed by SRI International in the 1980s. Like Kevlar, Zylon is used in a number of applications that require very high strength with excellent thermal stability. Tennis racquets, table tennis blades, various medical applications, and some of the Martian rovers are some of the more well known instances.

IUPAC name: poly(p-phenylene-2,6-benzobisoxazole)

Usage

Body armor : Zylon gained wide use in U.S. police officers body armor protection in 1998 with its introduction by Second Chance Body Armor, Inc. But protective vests constructed with Zylon became controversial in late 2003 when Oceanside, CA Police Officer Tony Zeppetella's and Forest Hills, PA Police Officer Ed Limbacher's vests failed, leaving Zeppetella mortally wounded and Limbacher seriously injured. Some studies subsequently reported that the Zylon vests may degrade rapidly, leaving wearers with significantly less protection than expected. Second Chance eventually recalled all of its zylon-containing vests, which led to its subsequent bankruptcy. In early 2005, Armor Holdings, Inc. first recalled its existing Zylon-based products, and decreased the rated lifespan warranty of new vests from 60 months to 30 months. In August 2005, AHI decided to discontinue manufacturing all of its Zylon-containing vests. This was largely based on the actions of the U.S. government's National Institute of Justice, which decertified Zylon for use in its approved models of ballistic vests for law enforcement.

Space Elevator Research : A competition was held in the Wirefly X Prize Cup in Las Cruces, New Mexico, US, on October 20-21, 2006. A team from the University of British Columbia entered into the Tether Challenge, using a construction made from Zylon fibres. And the house tether used by Spaceward that the other teams would have to beat in strength by 50% in the 2007 strong tether challenge was made out of Zylon.

Formula One Racing : As has been the case since 2001, Zylon tethers are used in F1 to affix the wheel to the chassis thus preventing the wheel from ejecting in to a crowded area in the event of an accident which causes the wheel to become air borne. Starting in the 2007 season, the driver's cockpit must now be clad in special anti-penetration panels made of Zylon. The Indy Racing League will also use Zylon starting in 2008.

Standing rigging : On modern racing yachts Zylon is used for parts of the standing rigging. It is used as shrouds and stays except the forestay due to torsion weakening the fibre. The PBO fibre degrade by UV and visible light, seawater and shafing and is therefore protected by a synthetic melted-on jacket. It is claimed to be 65% lighter than traditional rigging and the price 110-130% of rodrigging. And even the durability is claimed to be superior, but the statement is based on laboratory tests. It also has a high Young's modulus of 270GPa, meaning that it is very stiff.

Vectran

Vectran is a manufactured fibre, spun from a liquid crystal polymer created by Celanese Acetate LLC and now manufactured by Kuraray Co., Ltd. Chemically it is an aromatic polyester.

Properties: These fibres are noted for thermal stability at high temperatures, high strength and modulus, low creep, and good chemical stability. They are moisture resistant and are generally stable in hostile environments. They have gold color. They are often used in combination with some polyester as a coating around Vectran core; polyurethane coating can improve abrasion resistance and act as a water barrier. Vectran has melting point of 330 °C, with progressive strength loss from 220 °C. As it has high resistance to ultraviolet radiation, it can be used outside for long term, if inspected regularly.

Usage

They are used as reinforcing (matrix) fibres for ropes, cables, and advanced composite materials, professional bike tires, and in electronics applications. Perhaps most notably, Vectran is used as one of the five layers in NASA's current space suit design, and was the fabric used for the airbags on the Mars Pathfinder and twin Mars Exploration Rovers *Spirit* and *Opportunity* missions, which allowed those craft to "soft land" on Mars in 1997 and 2004, respectively. The material is expected to be used again on NASA's 2011 Mars Science Laboratory in the bridle cables. It is also used in the inflatable spacecraft developed by Bigelow Aerospace, the first of which (*Genesis I*) has been in space since July 12, 2006.

Production : Kuraray Co., Ltd. owns 100% of the worldwide Vectran production since 2005 as they acquired the Vectran business from Celanese Advanced Materials Inc. (CAMI), based in South Carolina, U.S.

In 2007 the total capacity of Vectran is about 600 tons/yr, but expanded to 1000 tons/yr in 2008.

Acrylonitrile

Acrylonitrile is the chemical compound with the formula CH_2CHCN. This pungent-smelling colorless liquid often appears yellow due to impurities. It is an important monomer for the manufacture of useful plastics. In terms of its molecular structure, it consists of a vinyl group linked to a nitrile.

Uses : Acrylonitrile is used principally as a monomer in the manufacture of synthetic polymers, especially polyacrylonitrile which comprises acrylic fibres. Acrylic fibres are, among other uses, precursor for well-known carbon-fibre. It is also a component of synthetic rubber. Dimerization of acrylonitrile affords adiponitrile, used in the synthesis of certain Nylons. Small amounts are also used as a fumigant. Acrylonitrile and derivatives such as **2-chloro-acrylonitrile** are dienophiles in Diels-Alder reactions. Acrylonitrile is also a precursor in the industrial manufacture of acrylamide and acrylic acid.

Production : Acrylonitrile is probably the nitrile manufactured on the largest scale. Most industrial acrylonitrile is produced through the Sohio process, the catalytic ammoxidation of propylene:

$2CH_3$-CH=CH_2 + $2NH_3$ + $3O_2$ → $2CH_2$=CH-Ca"N + $6H_2O$

World annual production in 2001 was 4 million tonnes and in 2005 6 million tonnes, with just under half of that coming from the United States. A significant by product is acetonitrile; this process is the main source of acetonitrile.

In 1996 the Sohio acrylonitrile process was designated an ACS National Historical Chemical Landmark in recognition of the innovative single-step method of production that made acrylonitrile available for chemical manufacturing worldwide.

Health Effects : Acrylonitrile is highly flammable and toxic. It undergoes explosive polymerization. The burning material releases fumes of hydrogen cyanide and oxides of nitrogen. The International Agency for Research on Cancer (IARC) concluded that there is inadequate evidence in humans for the carcinogenicity of acrylonitrile. [1] Acrylonitile increases cancer in high dose tests in male and female rats and mice.

Fibreglass

Fibre glass, (also called **fibre glass** and **glass fibre**), is material made from extremely fine fibres of glass. It is used as a reinforcing agent for many polymer products; the resulting composite material, properly known as fibre-reinforced polymer (FRP) or glass-reinforced plastic (GRP), is called "fibreglass" in popular usage. Glass makers throughout history have experimented with glass fibres, but mass manufacture of fibreglass was only made possible with the invention of finer machine tooling. In 1893, Edward Drummond Libbey exhibited a dress at the World's Columbian Exposition incorporating glass fibres with the diameter and texture of silk fibres. This was first worn by the popular stage actress of the time Georgia Cayvan.

What is commonly known as "fibreglass" today, however, was invented in 1938 by Russell Games Slayter of Owens-Corning as a material to be used as insulation. It is marketed under the trade name **Fibreglas,** which has become a genericized trademark. A somewhat similar, but more expensive technology used for applications requiring very high strength and low weight is the use of carbon fibre.

Formation

Glass fibre is formed when thin strands of silica-based or other formulation glass is extruded into many fibres with small diameters suitable for textile processing. Glass, even as a fibre, has little crystalline structure. The properties of the structure of glass in its softened stage are very much like its properties when spun into fibre. One definition of glass is "an inorganic substance in a condition which is continuous with, and analogous to the liquid state of that substance, but which, as a result of a reversible change in viscosity during cooling, has attained so high a degree of viscosity as to be, for all practical purposes, rigid."

The technique of heating and drawing glass into fine fibres has been known for millennia; however, the use of these fibres for textile applications is more recent. The first commercial production of fibreglass was in 1936. In 1938, Owens-Illinois Glass Company and Corning Glass Works joined to form the Owens-Corning Fibreglas Corporation.

Until this time all fibreglass had been manufactured as staple. When the two companies joined to produce and promote fibreglass, they introduced continuous filament glass fibres. Owens-Corning is still the major fibreglass producer in the market today. Two types of fibreglass most commonly used are S-glass and E-glass. E-glass has good insulation properties and it will maintain its properties up to 1500 degree F(815 deg C). S-glass has a high tensile strength and is stiffer than E-glass.

Chemistry

The basis of textile-grade glass fibres is silica, SiO_2. In its pure form it exists as a polymer, $(SiO_2)_n$. It has no true melting point but softens at 2000°C, where it starts to degrade. At 1713°C, most of the molecules can move about freely. If the glass is then cooled quickly, they will be unable to form an ordered structure. In the polymer, it forms SiO_4 groups which are configured as a tetrahedron with the silicon atom at the center and four oxygen atoms at the corners. These atoms then form a network bonded at the corners by sharing the oxygen atoms.

The vitreous and crystalline states of silica (glass and quartz) have similar energy levels on a molecular basis, also implying that the glassy form is extremely stable. In order to induce crystallization, it must be heated to temperatures above 1200°C for long periods of time.

Although pure silica is a perfectly viable glass and glass fibre, it must be worked with at very high temperatures, which is a drawback unless its specific chemical properties are needed. It is usual to introduce impurities into the glass in the form of other materials to lower its working temperature. These materials also impart various other properties to the glass which may be beneficial in different applications. The first type of glass used for fibre was soda lime glass or A glass. It was not very resistant to alkali. A new type, E-glass was formed that is alkali free (< 2%) and is an alumino-borosilicate glass. This was the first glass formulation used for continuous filament formation. E-glass still makes up most of the fibreglass production in the world. Its particular components may differ slightly in percentage, but must fall within a specific range. The letter E is used because it was originally for electrical applications. S-glass is a high-strength formulation for use when tensile strength is the most important property. C-glass was developed to resist attack from chemicals, mostly acids which destroy E-glass. T-glass is a North American variant of C-glass. A-glass is an industry term for cullet glass, often bottles, made into fibre. AR-glass is alkali-resistant glass. Most glass fibres have limited solubility in water but are very dependent on pH. Chloride ions will also attack and dissolve E-glass surfaces. A recent trend in the industry is to reduce or eliminate the boron content in the glass fibres.

Since E-glass does not really melt, but soften, the softening point is defined as "the temperature at which a 0.55 – 0.77 mm diameter fibre 235 mm long, elongates under its own weight at 1 mm/min when suspended vertically and heated at the rate of 5°C per minute". The strain point is reached when the glass has a viscosity of $10^{14.5}$ poise. The annealing point, which is the temperature where the internal stresses are reduced to an acceptable commercial limit in 15 minutes, is marked by a viscosity of 10^{13} poise.

Properties : Glass fibres are useful because of their high ratio of surface area to weight. However, the increased surface area makes them much more susceptible to chemical attack.

By trapping air within them, blocks of glass fibre make good thermal insulation, with a thermal conductivity on the order of 0.05 W/(mK).

Glass strengths are usually tested and reported for "virgin" fibres: those which have just been manufactured. The freshest, thinnest fibres are the strongest because the thinner fibres are more ductile. The more the surface is scratched, the less the resulting tenacity. Because glass has an amorphous structure, its properties are the same along the fibre and across the fibre. Humidity is an important factor in the tensile strength. Moisture is easily adsorbed, and can worsen microscopic cracks and surface defects, and lessen tenacity.

In contrast to carbon fibre, glass can undergo more elongation before it breaks. There is a correlation between bending diameter of the filament and the filament diameter. The viscosity of the molten glass is very important for manufacturing success. During drawing (pulling of the glass to reduce fibre circumference), the viscosity should be relatively low. If it is too high, the fibre will break during drawing. However, if it is too low, the glass will form droplets rather than drawing out into fibre.

Glass-reinforced Plastic

Glass-reinforced plastic (GRP) is a composite material or fibre-reinforced plastic made of a plastic reinforced by fine glass fibres. Like graphite-reinforced plastic, the composite material is commonly referred to by the name of its reinforcing fibres (fibreglass). Chemosetting plastics are normally used for GRP production—most often polyester (using butanone as a catalyst), but vinylester or epoxy are also used. The glass can be in the form of a chopped strand mat (CSM) or a woven fabric.

As with many other composite materials (such as reinforced concrete), the two materials act together, each overcoming the deficits of the other. Whereas the plastic resins are strong in compressive loading and relatively weak in tensile strength, the glass fibres are very strong in tension but have no strength against compression. By combining the two materials, GRP becomes a material that resists both compressive and tensile forces well. The two materials may be used uniformly or the glass may be specifically placed in those portions of the structure that will experience tensile loads.

Uses : Uses for regular fibreglass include mats, thermal insulation, electrical insulation, reinforcement of various materials, tent poles, sound absorption, heat- and corrosion-resistant fabrics, high-strength fabrics, arrows, bows and crossbows, translucent roofing panels, automobile bodies, electrical insulation and boat hulls.

Metallic Fibres

Metallic fibres are manufactured fibres composed of metal, plastic-coated metal, metal-coated plastic, or a core completely covered by metal. Gold and silver have been used since ancient times as yarns for fabric decoration. More recently, aluminum yarns, aluminized plastic yarns, and aluminized nylon yarns have replaced gold and silver. Metallic filaments can be coated with transparent films to minimize tarnishing. A common film is Lurex polyester.

History : Gold and silver have been used since ancient times as decoration in the clothing and textiles of kings, leaders, nobility and people of status. Many of these elegant textiles can be found in museums around the world. Historically, the metallic thread was constructed by wrapping a metal strip around a fibre core (cotton or silk), often in such a way as to reveal the color of the fibre core to enhance visual quality of the decoration. Ancient textiles and clothing woven from wholly or partly gold threads is sometimes referred to as *Cloth of Gold*. They have been woven on Byzantine looms from the 7th to 9th Centuries, and after that in Sicily, Cyprus, Lucca, and Venice. Weaving also flourished in the 12th Century during the legacy of Genghis Khan when art and trade flourished under Mongol rule in China and some Middle Eastern areas. The Dobeckmum Company produced the first modern metallic fibre in 1946.[3] In the past, aluminum was usually the base in a metallic fibre. More recently stainless steel has become a base as well. It is more difficult to work with but provides properties to the yarn that allows it to be used in more high tech applications.

Fibre properties : Coated metallic filaments help to minimize tarnishing. When suitable adhesives and films are used, they are not affected by salt water, chlorinated water in swimming pools or climatic conditions. If possible anything made with metallic fibres should be dry-cleaned, if there is no care label. Ironing can be problematic because the heat from the iron, especially at high tempatures, can melt the fibres.

Production method; There are two basic processes that are used in manufacturing metallic fibres. The most common is the *laminating process*, which seals a layer of aluminum between two layers of acetate or polyester film. These fibres are then cut into lengthwise strips for yarns and wound onto bobbins. The metal can be colored and sealed in a clear film, the adhesive can be colored, or the film can be colored before laminating. There are many different variations of color and effect that can be made in metallic fibres, producing a wide range of looks.

Metallic fibres can also be made by using the *metalizing process*. This process involves heating the metal until it vaporizes then depositing it at a high pressure onto the polyester film. This process produces thinner, more flexible, more durable, and more comfortable fibres.

Producers: Currently metallic fibres are manufactured primarily in Europe with only three manufacturers still producing metallic yarn in the United States. Metlon Corporation is one of the remaining manufacturers in the U.S. that stocks a wide variety of laminated and non-laminated metallic yarns.

Trademarks : The Lurex Company has manufactured metallic fibres in Europe for over fifty years. They produce a wide variety of metallic fibre products including fibres used in apparel fabric, embroidery, braids, knitting, military regalia, trimmings, ropes, cords, and lace surface decoration. The majority of Lurex fibres have a polyamdie film covering the metal strand but polyester and viscose are also used. The fibres are also treated with a lubricant called P.W., a mineral-based oil, which helps provide ease of use.

Metlon Corporation is a trademark of Metallic Yarns in the United States and has been producing metallic yarns for over sixty years. Metlon produces their metallic yarn by wrapping single slit yarns with two ends of nylon. One end of nylon is wrapped clockwise and the other end is wrapped counterclockwise around the metallic yarn. The most commonly used nylon is either 15 denier or 20 denier, but heavier deniers are used for special purposes.

Uses

The most common uses for metallic fibres is upholstery fabric and textiles such as lamé and brocade. Many people also use metallic fibres in weaving and needlepoint. Increasingly common today are metaillic fibres in clothing, anything from party and evening wear to club clothing, cold weather and survival clothing, and everyday wear. Metallic yarns are woven, braided, and knit into many fashionable fabrics and trims. For additional variety, metallic yarns are twisted with other fibres such as wool, nylon, cotton, and synthetic blends to produce yarns which add novelty effects to the end cloth or trim. Stainless steel and other metal fibres are used in communication lines such as phone lines and cable television lines. Stainless steel fibres are also used in carpets. They are dispersed throughout the carpet with other fibres so they are not detected. The presence of the fibres helps to conduct electricity so that the static shock is reduced. These types of carpets are often used in computer-use areas where the chance of producing static is much greater. Other uses include tire cord, missile nose cones, work clothing such as protective suits, space suits, and cut resistant gloves for butchers and other people working near bladed or dangerous machinery.

Bibliography

Philip SYKAS, Textile History 32, 2001, p. 156
David M. SWAN, Textile History 32, 2001, p. 175
Jennie BRUNTON, Textile History 32, 2001, p. 217
Steven KING; Textile History 33, 2002, p. 1
John STYLES, Textile History 33, 2002, p. 9
Sam SMILES, Textile History 33, 2002, p. 22
Steven KING, Textile History 33, 2002, p. 37
Christiana PAYNE, Textile History 33, 2002, p. 48
Christine STEVENS, Textile History 33, 2002, p. 63
Brian MAIDMENT, Textile History 33, 2002, p. 79
C.M. JACKSON-HOULSTON, Textile History 33, 2002, p. 98
Karen SAYER, Textile History 33, 2002, p. 112

Roy, Nilima, Art of Manipur, Agam Kala Prakashan, New Delhi, 1979.

Ganguli, Milada, Naga Art, New Delhi, Oxford & IBH

The Arts and Crafts of Nagaland, Naga institute of Culture, Government of Nagaland, Kohima, 1968.

Bhavnani, Enakshi, Folk and Tribal designs of India, Taraporevala, Bombay, 1974.

Barpujari, H.K. The comprehensive History of Assam. Vol V, Publication board of Assam, 1993.

Nyori, Tai, History and Culture of the Adis, Omsons Publications, 1993.

Panchani, Chandra Sheikhar, Arunachal Pradesh: Religion, Culture & society, Konark Publishers, New Delhi, 1989.

Panchani, Chandra Sheikhar, Manipur: Religion, Culture & society, Konark Publishers, New Delhi, 1987.

Chattopadhyay, Kamaladevi, The glory of Indian handicrafts, Indian book Company, Delhi, 1976.

Saraf, D.N. Indian Crafts- Development and Potential, Vikas Publishing House, Delhi, 1982.

Saraf, D.N. In the journey of Crafts development, (1941- 1991), Sampark Publishing house, Delhi, 1992.

Elwin Verrier, The Art of Northeast Frontier of India, North-east frontier agency, Shillong, 1959.

Hussain, Majid, Encyclopedia of India: Assam, Rima Publications, 1994.

David M. HIGGINS, Textile History 31, 2000, p. 54

Sue BOWDEN; Paul TURNER, Textile History 31, 2000, p. 77

J.I. GREAVES, Textile History 31, 2000, p. 102

Harald DECEULAER, Textile History 31, 2000, p. 133

Paul H. REM, Textile History 31, 2000, p. 150

Winifred ALDRICH, Textile History 31, 2000, p. 163

Andy BIELENBERG, Textile History 31, 2000, p. 202

Ian MORTIMER; Joseph MELLING, Textile History 31, 2000, p. 222

Anna MUTHESIUS, Textile History 32, 2001, p. 1

Lisa MONNAS, Textile History 32, 2001, p. 2

Anna MUTHESIUS, Textile History 32, 2001, p. 36

Frances PRITCHARD, Textile History 32, 2001, p. 48

Marianne ELLIS, Textile History 32, 2001, p. 61

Petra LINCHEID, Textile History 32, 2001, p. 75

Alisa BAGINSKI, Textile History 32, 2001, p. 81

Orit SHAMIR, Textile History 32, 2001, p. 93

Linda WOOLLEY, Textile History 32, 2001, p. 106

Marguerite KNOCKERT, Textile History 32, 2001, p. 114

Philip A. *SYKAS*, Textile History 32, 2001, p. 119

Patricia WARDLE, Textile History 32, 2001, p. 133

Hussain, Majid, Encyclopedia of Northeast India: Arunachal Pradesh, Vol. I, Rima Publications, 1994.

Hussain, Majid, Encyclopedia of India: Nagaland, Rima Publications, 1994.

Crill, Rosemary, 'Vaishnavite Silks: The figured textiles of Assam' in The woven silks of Assam, Marg, Publication, 1998.

Dhamija Jasleen, 'The Survey of Embroidery Traditions' in Textiles and Embroideries of India, Mark Publications, Bombay, 1965.

Iyer D.S.V. 'Looms', in Textiles and Embroideries of India, Mark Publications, Bombay, 1965.

Mudaliar, M.P.Nachimuthu, 'Manipur, Naga and Assam Fabrics', in Textiles and Embroideries of India, Mark Publications, Bombay, 1965.

Hussain, Majid, N.S.Olaniya, Encyclopedia of India: Arunachal Pradesh & Mizoram, Rima Publications, 1994.

Ali, A. Yusuf, A monograph of Silk Fabrics, Allahbad, 1900.

Indian textiles in the 20th century: Crisis in transformation, in 2000: Reflections on the arts of India, Marg publications, 2000.

Barve, V.R. Complete textile encyclopedia, Russi Lal Publishers, Bombay, 1967.

Index

A

B

C